T0259701

SpringerBriefs in Energy

More information about this series at http://www.springer.com/series/8903

Pratima Bajpai

Third Generation Biofuels

 Springer

Pratima Bajpai
Pulp and Paper Consultant
Kanpur, Uttar Pradesh, India

ISSN 2191-5520   ISSN 2191-5539   (electronic)
SpringerBriefs in Energy
ISBN 978-981-13-2377-5   ISBN 978-981-13-2378-2   (eBook)
https://doi.org/10.1007/978-981-13-2378-2

Library of Congress Control Number: 2018957696

This Springer imprint is published by the registered company Springer Nature Singapore Pte Ltd.
The registered company address is: 152 Beach Road, #21-01/04 Gateway East, Singapore 189721,
Singapore

Preface

The growing concerns about rapid depletion of fossil fuel reserves, increasing crude oil price, energy security and global climate change have led to growing worldwide interests in renewable energy sources such as biofuels. The biofuel production from renewable sources is widely considered to be one of the most sustainable alternatives to fossil fuels and a viable means for environmental and economic sustainability. An increasing number of developed and rapidly developing nations see biofuels as a key to reducing reliance on foreign oil, lowering emissions of greenhouse gases, mainly carbon dioxide and methane, and meeting rural development goals. Biofuels are referred to solid, liquid or gaseous fuels derived from organic matter. They are generally divided into primary and secondary biofuels. While primary biofuels such as fuelwood are used in an unprocessed form primarily for heating, cooking or electricity production, secondary biofuels such as bioethanol and biodiesel are produced by processing biomass and are able to be used in vehicles and various industrial processes. The secondary biofuels can be categorized into three generations: first, second and third generation biofuels on the basis of different parameters, such as the type of processing technology, type of feedstock or their level of development.

Although biofuel processes have a great potential to provide a carbon-neutral route to fuel production, first generation production systems have considerable economic and environmental limitations. The most common concern related to the current first generation biofuels is that as production capacities increase, so does their competition with agriculture for arable land used for food production. The increased pressure on arable land currently used for food production can lead to severe food shortages, in particular for the developing world where already more than 800 million people suffer from hunger and malnutrition. In addition, the intensive use of land with high fertilizer and pesticide applications and water use can cause significant environmental problems. The advent of second generation biofuels is intended to produce fuels from lignocellulosic biomass, the woody part of plants that do not compete with food production. Sources include agricultural residues, forest harvesting residues or wood processing waste such as leaves, straw or wood chips as well as the non-edible components of corn or sugarcane. However, converting the woody biomass into fermentable sugars requires costly technologies involving pre-treatment with special enzymes, meaning that second generation biofuels cannot yet be produced economically on a large scale. Therefore, third generation biofuels derived from microalgae are considered to be a viable alternative energy resource that is devoid of the major drawbacks associated with first and second generation biofuels. Microalgae are able to produce 15–300 times more oil for biodiesel production than traditional crops on an area basis. Furthermore compared with conventional crop plants which are usually harvested once or twice a year, microalgae have a very short harvesting cycle allowing multiple or continuous harvests with significantly increased yields. This book examines the background to third generation biofuel production, advantages of algae over traditional biofuel crops, algal biomass production, algae harvesting and drying methods, production of biofuel from microalgae and future prospects. Some of the leading companies involved with third generation biofuel research and development are also presented.

Kanpur, Uttar Pradesh, India   Pratima Bajpai

About the Book

Growing concerns about the rapid depletion of fossil fuel reserves, rising crude oil prices, energy security and global climate change have led to increased worldwide interest in renewable energy sources such as biofuels. In this context, biofuel production from renewable sources is considered to be one of the most sustainable alternatives to fossil fuels and a viable means of achieving environmental and economic sustainability.

Although biofuel processes hold great potential to provide a carbon-neutral route to fuel production, first-generation production systems are characterized by considerable economic and environmental limitations. The advent of second-generation biofuels is intended to produce fuels from lignocellulosic biomass, the woody part of plants that does not compete with food production. However, converting woody biomass into fermentable sugars requires costly technologies. Therefore, third-generation biofuels from microalgae are considered to be a viable alternative energy resource, free from the major drawbacks associated with first- and second-generation biofuels. This book examines the background of third-generation biofuel production, the advantages of algae over traditional biofuel crops, algal biomass production, algae harvesting and drying methods, production of biofuel from microalgae, and future prospects.

Contents

About the Author

Dr. Pratima Bajpai is currently working as a Consultant in the field of Paper and Pulp. She received her Ph.D. from the National Sugar Institute, Kanpur, India. She has worked at the University of Saskatchewan (Canada), University of Western Ontario (Canada), and Thapar Center for Industrial Research and Development (India). She has also worked as a visiting Professor at the University of Waterloo, Canada, and visiting researcher at Kyushu University, Fukuoka, Japan. Dr. Bajpai's main areas of expertise are industrial biotechnology, paper and pulp, and environmental biotechnology. A recognized expert in her field, she has more than 150 publications in international journals and conference proceedings to her credit. She has written several advanced-level technical books for leading global publishers, has contributed to a number of chapters in books and encyclopedias, and holds 11 patents.

List of Figures

List of Tables

Chapter 1
General Background and Introduction

Abstract Fossil fuel sources are getting exhausted and contribute to greenhouse gas (GHG) emissions which leads to many negative effects. Hence, biofuels are being explored to replace fossil fuels. Biofuels are favourable choice of fuel consumption due to their renewability, biodegradability and generating acceptable quality exhaust gases. Third generation biofuels derived from microalgae are considered to be a viable alternative energy resource devoid of the major drawbacks associated with first and second generation biofuels. This chapter presents general background and introduction on third generation biofuels from algae.

Keywords Fossil fuel · Greenhouse gas · Biofuels · Microalgae · Third generation biofuels · Biomass · Photosynthetic microorganisms

Fossil fuels are major energy source in the world and constitute more than 80% of the worldwide energy supply (www.studentenergy.org/topics/fossil-fuels). Non-OECD countries are holding majority of the proven reserves for all fossil fuels (Singh and Rathore 2017). These energy sources are powering, the industrialization of nations (www.studentenergy.org/topics/fossil-fuels). They have several applications which range from production of electricity to transport fuel. Fossil fuels have been also explored for polymers production including plastics, paints, detergents, cosmetics and certain medicines (www.studentenergy.org/topics/fossil-fuels). Some fossil fuels like coal, are available in abundance and are an inexpensive source of energy. Others, such as oil, have a variable cost depending on geographic location. Geopolitical issues arise because of this reason, due to the geographic allocation of these highly valuable resources (www.studentenergy.org/topics/fossil-fuels). Fossil fuels have taken millions of years to get formed and are non-renewable resources. Fossil fuel sources are getting exhausted and contribute to emissions of green house gases leading to several unfavourable effects which includes (internationalscienceindex.org):

- change in climate
- receding of glaciers
- increase in sea level
- loss of biodiversity

© The Author(s), under exclusive license to Springer Nature Singapore Pte Ltd. 2019
P. Bajpai, *Third Generation Biofuels*, SpringerBriefs in Energy,
https://doi.org/10.1007/978-981-13-2378-2_1

The increase in energy demand results in an increase in the price of crude oil which has an effect on the economic activity worldwide. Progressive exhaustion of conventional fossil fuels with increasing consumption of energy along with emissions of GHG have led to a move towards efficient, sustainable, renewable, and profitable sources of energy with reduced emissions (www.frontiersin.org; Dragone et al. 2010; Rajkumar et al. 2014; Goldemberg and Guardabassi 2009). Therefore, biofuels are being examined for replacing fossil fuels. Biofuels are, renewable, biodegradable and are able to produce acceptable quality exhaust gases (xa.yimg.com; Dragone et al. 2010; Brennan and Owende 2010; Scott et al. 2010; Mobin & Chowdhury 2015; Escobar et al. 2009). Biofuels remain the most environment friendly and practical solution to the global fuel crisis.

Biomass from different sources such as forestry, agricultural, and aquatic have been examined as substrates for biofuel production (www.frontiersin.org). Biofuels are divided into two categories – primary biofuels and secondary biofuels (repositorium.sdum.uminho.pt) (Table 1.1). The primary biofuels are not processed and are used for heating, electricity production or cooking. Secondary biofuels such as bioethanol and biodiesl are produced from biomass. These are used in several industrial processes and in vehicles. Depending upon the type of processing technology, type of substrate, the secondary biofuels can be classified into following categories (Dragone et al. 2010):

- first generation biofuel
- second generation biofuel
- third generation biofuel

Table 1.1 Biofuels

Primary biofuel
Firewood, wood chips, pellets, animal waste, forest and crop residues, landfill gas
Secondary biofuel
1st generation
Bioethanol or butanol by fermentation of starch (from wheat, barley, corn, potato) or sugars (from sugarcane, sugar beet, etc.)
Biodiesel by transesterification of oil crops (rapeseed, soybeans, sunflower, palm, coconut, used cooking oil, animal fats, etc.)
2nd generation
Bioethanol and biodiesel produced from conventional technologies but based on novel starch, oil and sugar crops such as Jatropha, cassava or Miscanthus;
Bioethanol, biobutanol, syndiesel produced from lignocellulosic materials (e.g. straw, wood, and grass)
3rd generation
Biodiesel from microalgae
Bioethanol from microalgae and seaweeds
Hydrogen from green microalgae and microbes

Based on Alam et al. (2012), Yanqun et al. (2008), Behera et al. (2014), Boyce et al. (2008), Marques et al. (2011), Shuping et al. (2010), Hughes et al. (2012), Singh et al. (2011, 2014)

"Biofuel processes provide a carbon-neutral route to fuel production. The first generation fuels however, have significant environmental and economic limitations. One of the most common issue related to the first generation biofuels is that with the increase of production capacities, the competition with agriculture for arable land used for production of food also increases" (Dragone et al. 2010; repositorium. sdum.uminho.pt). The increased pressure on land presently used for production of food can lead to acute food shortages particularly for the developing countries where many people suffer from malnutrition and hunger. Furthermore, exhaustive use of land with large amount of fertilizer, pesticide and water can cause severe environmental issues. Lignocellulosic biomass do not compete with production of food. These are used for production of second generation biofuels (agronomy.emu. ee). Sources include wood processing waste like wood chips, leaves or straw, the non-edible portions of corn or sugarcane, agricultural and forest harvesting residues (repositorium.sdum.uminho.pt). But, conversion of the woody material into fermentable sugars uses expensive techniques such as enzymatic pre-treatment (Alam et al. 2012; repositorium.sdum.uminho.pt). This shows that producing second generation biofuels on a large scale is not yet economically viable. Third generation biofuels from microalgae lack the main problems associated with first and second generation biofuels (repositorium.sdum.uminho.pt). Therefore, these biofuels appear to be a viable alternative energy resource. Microalgae on an area basis produce more oil for production of biodiesel in comparison to the traditional crops (cest2015.gnest.org). The harvesting cycle in microalgae is very short. This allows continuous harvests and provide substantially higher yields. In case of conventional crops, harvesting is done once or twice a year (veprints.unica.it). According to Demirbas (2007), algal biomass does not compete with food and feed production (www.frontiersin.org). Microalgae are photosynthetic microorganisms and need light, carbon dioxide and nutrients for growth, and for producing carbohydrates, which can be processed into biofuels and other value added products (www.frontiersin.org; Brennan and Owende 2010; Nigam and Singh 2011). The nutrients required are nitrogen, phosphorus, and potassium. In algal biomass, the hemicellulose content is very low and the lignin content is zero. This increases the hydrolysis rate and fermentation efficiency (Saqib et al. 2013). Algae are used in several other areas such as animal feed, human nutrition, pharmaceuticals, nutraceuticals, and cosmetics, biofertilizer and bioremediation (Choi et al. 2012; Hsueh et al. 2007; Tamer et al. 2006; Crutzen et al. 2007; Thomas 2002). Growing of microalgae also can complement approaches like bioremediation of wastewaters and carbon dioxide sequestration thus addressing the serious environmental concerns.

Figure 1.1 shows biofuel production from microalgae. Figure 1.2 shows production of bioethanol and biodiesel using microalgae as a substrate.

Algae appears to be a promising substrate for production of second- and third-generation biofuels. Several new companies are investing in this project. (https://snrec-mitigation.wordpress.com/2009/03/12/third-generation-biofuels/). Algae also show several advantages over traditional crops and holds promise for production of biofuels.

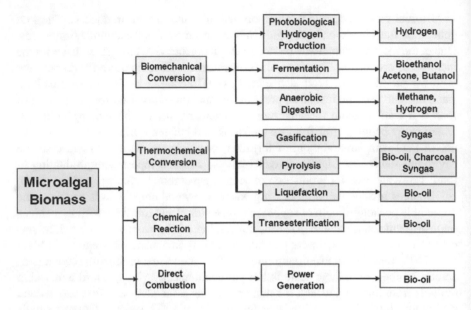

Fig. 1.1 Biofuel production processes from microalgae biomass. (Alam et al. 2012. Reproduced with permission)

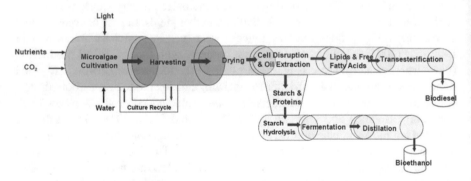

Fig. 1.2 Biodiesel and bioethanol production processes from microalgae. (Alam et al. 2012. Reproduced with permission)

Attempts are being made all over the world on development of biofuel from microalgae and several projects are getting implemented (Wang 2013). Several countries in the United States, Europe and Asia have their own microalgae research projects. Therefore, researchers are paying attention for developing microalgae for sustainable production of energy. Algae shows fast growth rate and has high content of oil. Production of algae requires less land and can be grown even in the shallow sea directly. Therefore, arable land can be saved for growing food. Compared to terrestrial crops, microalgae are more sustainable (Wang 2013).

Green algae, golden brown, prymnesiophytes, eustigmatophytes, cyanobacteria and Diatoms, have been studied as potential fuel production strains (Hannon et al. 2010). Cyanobacteria are a class of photosynthetic bacteria and are not algae.

References

Alam F, Date A, Rasjidin R, Mobin S, Moria H, Baqui A (2012) Biofuel from algae – is it a viable alternative? Procedia Eng 49:221–227

Behera S, Mohanty RC, Ray RC (2014) Batch ethanol production from cassava (Manihotesculenta Crantz.) flourusing Saccharomyces cerevisiae cells immobilized in calcium alginate. Ann Microbiol 65:779–783

Boyce AN, Chowd-Hury P, Naqiuddin M (2008) Biodiesel fuel production from algae a srenewable energy. Am J Biochem Biotechnol 4:250–254

Brennan L, Owende P (2010) Biofuels from microalgae – a review of technologies for production, processing, and extractions of biofuels and co-products. Renew Sust Energ Rev 14:557–577

Choi W, Han J, Lee C, Song C, Kim J, Seo Y (2012) Bioethanol production from Ulvapertusa kjellman by high-temperature liquefaction. Chem Biochem Eng 26:15–21

Crutzen PJ, Mosier AR, Smith KA, Winiwarter W (2007) N2O release fromagro-biofuel production negates global warming reduction by replacing fossil fuels. Atmos Chem Phys 7:11191–11205

Demirbas A (2007) Progress and recent trends in biofuels. Prog Energy Combust Sci 33:1–18

Dragone G, Fernandes B, Vicente AA, Teixeira JA (2010) Third generation biofuels from microalgae. In: Mendez-Vilas A (ed) Current research, technology and education topics in applied microbiology and microbial biotechnology. Formatex, Badajoz, pp 1355–1366

Escobar JC, Lora ES, Venturini OJ, Yanez EE, Castillo EF, Almazan O (2009) Biofuels: environment, technology and food security. Renew Sust Energ Rev 13:1275e87

Goldemberg J, Guardabassi P (2009) Are biofuels a feasible option? Energy Policy 37:10–14

Hannon M, Gimpel J, Tran M, Rasala B, Mayfield S (2010) Biofuels from algae: challenges and potential. Biofuels 1:763–784

Hsueh HT, Chu H, Yu ST (2007) A batch study on the bio-fixation of carbon dioxide in the absorbed solution from a chemical wet scrubber by hot spring and marine algae. Chemosphere 66:878–886

Hughes AD, Kelly MS, Black KD, Stanley MS (2012) Biogas from macroalgae: is it time to revisit the idea? Biotechnol Biofuels 5:1–7. https://doi.org/10.1186/1754-6834-5-86

Marques AE, Barbosa AT, Jotta J, Coelho MC, Tamagnini P, Gou-veia L (2011) Biohydrogen production by Anabaena sp.PCC7120 wild-type and mutants under different conditions: light, nickel, propane, carbon dioxide and nitrogen. Biomass Bioenergy 35:4426–4434

Mobin S, Chowdhury H (2015) Third generation biofuel from algae. Procedia Eng 105:763–768

Nigam PS, Singh A (2011) Production of liquid biofuels from renewable resources. Prog Energy Combust Sci 37(1):52–68

Rajkumar R, Yaakob Z, Takriff MS (2014) Potential of the micro and macroalgae for biofuel production: a brief review. Bioresources 9:1606–1633

Saqib A, Tabbssum MR, Rashid U, Ibrahim M, Gill SS, Mehmood MA (2013) Marine macroalgae Ulva: a potential feed-stock for bioethanol and biogas production. Asian J Agri Biol 1:155–163

Scott SA, Davey MP, Dennis JS, Horst I, Howe CJ, Lea Smith DJ, Smith AG (2010) Biodiesel from algae: challenges and prospects. Curr Opin Biotechnol 21:277–286

Shuping Z, Yulong W, Mingde Y, Kaleem I, Chun L, Tong J (2010) Production and characterization of bio-oil from hydrothermal liquefaction of micro algae Dunaliella tertiolecta cake. Energy 35:5406–5411

Singh A, Rathore D (2017) Biohydrogen production: sustainability of current technology and future perspective. Springer, New Delhi

Singh A, Olsen SI, Nigam PS (2011) A viable technology to generate thirdgeneration biofuel. J Chem Technol Biotechnol 86:1349–1353

Singh R, Behera S, Yadav YK, Kumar S (2014) Potential of wheat straw for biogas production using thermophiles. In: Kumar S, Sarma AK, Tyagi SK, Yadav YK (eds) Recent advances in bio-energy research. SSS- National Institute of Renewable Energy, Kapurthala, pp 242–249

Tamer E, Amin MA, Ossama ET, Bo M, Benoit G (2006) Biological treatment of industrial wastes in a photobioreactor. Water Sci Technol 53:117–125

Thomas DN (2002) Seaweeds. Smithsonian Books, Natural History Museum, Washington, DC

Wang Y (2013) Microalgae as the third generation biofuel: production, usage, challenges and prospects. Uppsala University, Uppsala

Yanqun L, Mark H, Nan W, Christopher QL, Nathalie DC (2008) Biofuels from microalgae. Biotechnol Prog 24:815–820

agronomy.emu.ee

cest2015.gnest.org

internationalscienceindex.org

link.springer.com

repositorium.sdum.uminho.pt

snrecmitigation.wordpress.com

veprints.unica.it

www.diva-portal.org

www.frontiersin.org

www.ncbi.nlm.nih.gov

www.studentenergy.org

xa.yimg.com

Chapter 2
Fuel Potential of Third Generation Biofuels

Abstract Microalgae are particularly suited for biofuel production due to their high photosynthetic growth rates, high lipid content, low land usage and high carbon dioxide absorption. Microalgae have a very short harvesting cycle (1–10 days depending on the process) compared with conventional crop plants which are usually harvested once or twice a year, allowing multiple or continuous harvests with significantly increased yields. Fuel potential of third generation biofuels from algae are presented in this chapter.

Keywords Biofuels · Microalgae · Third generation biofuels · Biomass · Photosynthetic microorganisms · Bioethanol · Biodiesel · Biogas · Biohydrogen

"No feedstock matches algae when it comes to the potential to produce fuel", (biofuel.org.uk). A variety of fuel can be produced from algae. This results mainly from the following characteristics of the microorganism:

- The algal oil can be used to produce diesel or gasoline and genetic modification can also be performed (slideplayer.com).

Butanol is alike to gasoline so it is of interest. Its energy density is similar to gasoline and has a better emission profile (biofuel.org.uk). Researchers faced problems in producing butanol until the development of genetically modified algae. Several commercial-scale plants have been set up and are producing butanol and other biofuels (biofuel.org.uk). Butanol does not cause any harm to the engine or require any change in engine like ethanol. Diversity is not the sole factor that algae has going for it in terms of fuel potential. From algae, several types of fuels can be produced. These are presented below (biofuel.org.uk):

- Ethanol
- Butanol
- Biodiesel
- Gasoline
- Methane

© The Author(s), under exclusive license to Springer Nature Singapore Pte Ltd. 2019
P. Bajpai, *Third Generation Biofuels*, SpringerBriefs in Energy,
https://doi.org/10.1007/978-981-13-2378-2_2

- Vegetable Oil
- Jet Fuel

Algae produces high yields. Per acre about 9000 gallons of biofuel can be produced. This is 10 times more than the conventional substrates (slideplayer.com). High yields upto 20,000 gallons per acre can also be obtained. Yields 10 times more in comparison to second generation biofuels shows that only 0.42% of the land in USA would be required for producing sufficient fuel for meeting the requirement in USA (biofuel.org.uk). The largest consumer of fuel in the world is in USA. Therefore, the efficiency of algal-based biofuels is obvious (biofuel.org.uk; slideplayer.com; link.springer.com; Silva and Ribeiro 2012).

Algae can use different types of carbon sources. This is one of the major benefits. Algae might be tied directly to carbon releasing sources (power plants, industry etc.) where they could directly convert emissions into fuel (biofuel.org.uk). Therefore, no carbon dioxide would be released from these settings and there would be a significant reduction in total emissions (biofuel.org.uk).

There are several thousands species of microalgae but not many species have been studied for production of biofuels (www.mdpi.com). Microalgae have high lipid content and the growth rate and absorption of carbon dioxide is high and the usage of land is low (www.mdpi.com; Singh and Olsen 2011). Algae is able to use sunlight and waste streams for providing important nutrients for growth. Waste carbon dioxide obtained from power plant exhaust gases can be converted by algae to biomass and then to energy (www.mdpi.com; Bruton et al. 2009). Furthermore, wastewater streams from municipality can be used for providing extra nutrients (Gusciute 2016; Singh et al. 2011). On an area basis, algae produces more oil (15–300 times) for production of biodiesel in comparison to the conventional crops (www.mdpi.com) (Table 2.1). The harvesting cycle is short in microalgae (1–10 days depending on the process) whereas in case of traditional crops, harvesting is done once or twice a year. This allows substantially higher yields (cest2015.gnest.org; Schenk et al. 2008).

Algae are both autotrophic and heterotrophic (journal.doi.org). The autotrophic algae needs light, carbon dioxide, and salts for their growth (Behera et al. 2015).

Table 2.1 Advantages of algal biomass for biofuels production

Ability to grow throughout the year, therefore, algal oil productivity is higher in comparison to the conventional oil seed crops
Higher tolerance to high carbon dioxide content
The consumption rate of water is very less in algae cultivation
No requirement of herbicides or pesticides in algal cultivation
The growth potential of algal species is very high in comparison to others
Different sources of wastewater containing nutrients like nitrogen and phosphorus can be utilized for algal cultivation apart from providing any additional nutrient
The ability to grow under harsh conditions like saline, brackish water, coastal sea water, which does not affect any conventional agriculture

Spolaore et al. (2006), Dismukes et al. (2008), Dragone et al. (2010), Ayyappan (2017)

The heterotrophs are non-photosynthetic, need nutrients and organic compounds (Brennan and Owende 2010). The size of microalgae is very small (in micrometers) and normally grow in ponds or water bodies. Lipid content in microalgae is more compared to macroalgae and the growth is fast (journal.doi.org; Lee et al. 2014). Several thousands of microalgal species have been researched (journal.doi.org; Rajkumar et al. 2014; Surendhiran and Vijay 2012; Richmond and Qiang 2013; Chisti 2007; Schenk et al. 2008). There are several advantages of algal biomass for biofuels production (journal.doi.org) (Table 2.1). Microalge are single-celled organisms. They multiply by division. Therefore, high-throughput techniques may be used for evolving the strains rapidly. The processes are reduced to a few months in algae; crop plants take years. Algae show reduced environmental impact in comparison with terrestrial biomass used for production of biofuels (Dismukes et al. 2008). The land which is not suitable for traditional agriculture can be used for growing algae. They can also remove nutrients from water efficiently (www.tandfonline.com). Algal biofuels would therefore reduce the use of land in comparison with biofuels produced from terrestrial plants. Waste streams (municipal wastewater; flue gas of coal) can also be treated during growing algae (www.tandfonline.com; He et al. 2008; Fierro et al. 2008; Douskova et al. 2009). Algae production strains can be bioengineered, and specific properties can be improved (Zaslavskaia et al. 2001; Rosenberg et al. 2008). Valuable co-products can be produced. This will make algal biofuels to compete with petroleum. "These characteristics make algae a platform with a high potential to produce biofuels in a cost effective manner" (www.tandfonline.com).

There are also downside of using microalgal biomass as substrate. The cost of cultivation is higher in comparison to traditional crops. The energy consumption for harvesting algae is also high (journal.doi.org). For harvesting and concentrating the algal biomass, different methods (flocculation, floatation, sedimentation, centrifugation and filtration) have been explored (Behera et al. 2015; Ho et al. 2011; Demirbas 2010).

Liquefaction and gasification routes are used for further processing for bio-oil and syngas production respectively (Behera et al. 2015; Algae and Environmental Sustainability 2015; journal.doi.org; Kraan 2013).

References

Ayyappan M (2017) Chapter 10: algal resource for sustainable food security. Springer, Singapore

Behera S, Singh S, Arora R, Sharma NK, Shukla M, Kumar S (2015) Scope of algae as third generation biofuels, Frontiers in bioengineering and biotechnology. Front Bioeng Biotechnol 2:90. Published online 2015 Feb 11. https://doi.org/10.3389/fbioe.2014.00090

Brennan L, Owende P (2010) Biofuels from microalgae – a review of technologies for production, processing, and extractions of biofuels and co-products. Renew Sust Energ Rev 14:557–577

Bruton T, Lyons H, Lerat Y, Stanley M, Rasmussen MB (2009) A review of the potential of marine algae as a source of biofuel in Ireland. Dublin, Sustainable Energy Ireland

Chisti Y (2007) Biodiesel from microalgae. Biotechnol Adv 25:294–306

Demirbas A (2010) Use of algae as biofuel sources. Energy Convers Manag 51:2738–2749

Dismukes GC, Carrieri D, Bennette N, Ananyev GM, Posewitz MC (2008) Aquatic phototrophs: efficient alternatives to land-based crops for biofuels. Curr Opin Biotechnol 19(3):235–240

Douskova I, Doucha J, Livansky K (2009) Simultaneous flue gas bioremediation and reduction of microalgal biomass production costs. Appl Microbiol Biotechnol 82(1):179–185

Dragone G, Fernandes B, Vicente AA, Teixeira JA (2010) Third generation biofuels from micro-algae. In: Mendez-Vilas A (ed) Current research, technology and education topics in applied microbiology and microbial biotechnology. Formatex, Badajoz, pp 1355–1366

Fierro S, Sanchez-Saavedra Mdel P, Copalcua C (2008) Nitrate and phosphate removal by chitosan immobilized Scenedesmus. Bioresour Technol 99(5):1274–1279

Gusciute E (2016) Transport sector in Ireland: can 2020 national policy targets drive indigenous biofuel production to success? In: Advances in bioenergy. Wiley, Oxford

He P, Xu S, Zhang H (2008) Bioremediation efficiency in the removal of dissolved inorganic nutrients by the red seaweed, Porphyra yezoensis, cultivated in the open sea. Water Res 42(4–5):1281–1289

Ho SH, Chen CY, Lee DJ, Chang JS (2011) Perspectives on microalgal CO_2-emission mitigation systems – a review. Biotechnol Adv 29:189–198

Kraan S (2013) Mass cultivation of carbohydrate rich microalgae, a possible solution for sustainable biofuel production. Mitig Adapt Strateg Glob Chang 18:27–46

Lee K, Eisterhold ML, Rindi F, Palanisami S, Nam PK (2014) Isolation and screening of micro-algae from natural habitats in the Midwestern United States of America for biomass and bio-diesel sources. J Nat Sci Biol Med 5:333–339

Rajkumar R, Yaakob Z, Takriff MS (2014) Potential of the micro and macroalgae for biofuel pro-duction: a brief review. Bioresources 9:1606–1633

Richmond A, Qiang H (2013) Handbook of microalgal culture: applied phycology and biotechnol-ogy, 2nd edn. Wiley-Blackwell, Hoboken

Rosenberg JN, Oyler GA, Wilkinson L, Betenbaugh MJ (2008) A green light for engineered algae: redirecting metabolism to fuel a biotechnology revolution. Curr Opin Biotechnol 19(5):430–436

Schenk PM, Thomas-Hall SR, Stephens E, Marx UC, Mussgnug JH, Posten C, Kruse O, Hankamer B (2008) Second generation biofuels: high-efficiency microalgae for biodiesel production. Bioenergy Res 1:20–43

Silva PP, And Ribeiro LA (2012). The role of microalgae in the deployment of biofuels: contrast-ing algae and solar technologies. Int J Technol Policy Manag;12: 158–176

Singh A, Olsen SI (2011) A critical review of biochemical conversion, sustainability and life cycle assessment of algal biofuels. Appl Energy 88:3548–3555

Singh A, Nigam PS, Murphy JD (2011) Mechanism and challenges in commercialisation of algal biofuels. Bioresour Technol 102:26–34

Singh B, Bauddh K, Bux F (eds) (2015) Algae and environmental sustainability. Springer, New Delhi

Spolaore P, Joannis-Cassan C, Duran E, Isambert A (2006) Commercial applications of microal-gae. J Biosci Bioeng 101:87–96

Surendhiran D, Vijay M (2012) Microalgal biodiesel-a comprehensive review on the potential and alternative biofuel. Res J Chem Sci 2:71–82

Zaslavskaia LA, Lippmeier JC, Shih C, Ehrhardt D, Grossman AR, Apt KE (2001) Trophic con-version of an obligate photoautotrophic organism through metabolic engineering. Science 292(5524):2073–2075

biofuel.org.uk

cest2015.gnest.org

journal.doi.org

link.springer.com

slideplayer.com

www.mdpi.com

www.tandfonline.com

Chapter 3
Characteristics of Algae

Abstract Algae are thallophytes and are a large, heterogeneous and polyphyletic group of simple plants and lack true roots, stem, leaves. Algae contain the pigment chlorophyll which is their primary photosynthetic pigment. The mechanism of photosynthesis in algae is similar to that of higher plants but they are generally more efficient converters of solar energy. Due to their simple cellular structure, they have more efficient access to water, carbon dioxide, and other nutrients because the cells grow in aqueous suspension. The characteristics of algae are presented in this chapter.

Keywords Microalgae · Macroalgae · Thallophytes · Pigment · Chlorophyll · Photosynthetic pigment · Photosynthesis · Autotrophic · Heterotrophic

Algae are of two types: macroalgae and microalgae. Macroalgae are multi-cellular algae, usually grow in ponds and the size is large. These algae grow in several ways and are measured in inches (pubs.ext.vt.edu). Seaweed are the largest multicellular algae; giant kelp plant is an example (articles.extension.org/pages/26600/algae-for-biofuel-production). Its length is more than 100 feet. Microalgae are unicellular and small in size. Microalgae is measured in micrometers and normally grow in suspension within water.

Algae are referred to as thallophytes as their plant body is called as thallus. These are large, heterogeneous and polyphyletic group of simple plants and lack true roots, stem, leaves (aepc.gov.np). Algae contain chlorophyll which is important for photosynthesis (www.formatex.info; Brennan and Owende 2010). The photosynthetic mechanism in algae resembles to that of higher plants. In algae, solar energy is utilized more efficiently. The algae grow in queous suspension and cellular structure is simple. They are able to access carbon dioxide, water and other nutrients efficiently (Chisti 2007; www.ret.gov.au; Geada et al. 2017).

Algae can be unicellular or multicellular (www.cliffsnotes.com/study...unicellular-algae/general-characteristics-of-algae). Unicellular algae are mainly found in water, particularly in plankton. Phytoplankton are composed mainly of unicellular algae and is the population of free-floating microorganisms (articles.directorym.com). Phytoplankton have chlorophyll and need sunlight to live and grow. They

P. Bajpai, *Third Generation Biofuels*, SpringerBriefs in Energy,
https://doi.org/10.1007/978-981-13-2378-2_3

resemble terrestrial plants. Most phytoplankton float in the upper part of the ocean, where sunlight passes in the water. Inorganic nutrients such as phosphates, sulfur and nitrates are also required by Phytoplankton. The nutrients are converted in to carbohydrates, fats and proteins. Two major classes of phytoplankton are dinoflagellates and diatoms. Dinoflagellates move through the water by using their flagella. Their bodies are covered with complex shells (articles.directorym.com). The structure of Diatoms is rigid. They have shells and made of interlocking parts. Diatoms do not depend on flagella to move through the water but depend on ocean currents for travelling through the water (www.waterboards.ca.gov; oceanservice.noaa.gov).

Algae may occur on the surface of wood, moist rocks or moist soil. Algae live with fungi in lichens. Algae can be classified into seven types, (each with distinct colour, size and functions) of which five are considered to be in the Protista kingdom and two in the Plantae kingdom (articles.directorym.com). The different divisions are presented below: (biology.about.com)

Euglenophyta
Chrysophyta
Chlorophyta
Xanthophyta
Rhodophyta
Paeophyta
Pyrrophyta

Algae possess eukaryotic properties. In some algal species, flagella are present showing the "9-plus-2" pattern of microtubules. A nucleus is present, and multiple chromosomes are seen in mitosis. Chloroplasts contain the chlorophyll and other pigments which contain thylakoids membranes (articles.directorym.com). Microalgae are autotrophic or heterotrophic (www.formatex.info). Inorganic compounds are used as a source of carbon in autotrophic algae. Autotrophs can be either photoautotrophic or chemoautotrophic. Photoautotrophs use light as a source of energy. Chemoheterotrophs derive energy from chemical reactions and nutrients from the organic matter. Heterotrophs can be either chemoheterotrophs or photoheterotrophs. For growth, heterotrophic algae utilize organic compounds (articles.directorym.com; www.dbbe.fcen.uba.ar; www.formatex.info).

Some microalgae are mixotrophic (use a mixture of different source of energy). These algae combine heterotrophy and autotrophy by photosynthesis (Lee 2008). Autotrophic algae, convert sun light and carbon dioxide into ATP (adenosine triphosphate) and oxygen, which is used for respiration for producing energy (Brennan and Owende 2010; www.formatex.info)

"Most species are saprobes and some are parasites. Saprobes derive their nourishment from dead or decaying organic matter. Reproduction occurs in both asexual and sexual forms in algae. Asexual reproduction takes place through the fragmentation of colonial and filamentous algae or by the formation of spores as in fungi. Spore formation takes place by mitosis. Binary fission also takes place as in case of bacteria. During sexual reproduction, algae form differentiated sex cells which fuse to form a diploid zygote which contains two sets of chromosomes. Sexual spore is formed from zygote. Under favorable conditions, this germinates to form the haploid organism having a single set of chromosomes. This type of reproduction pattern is called alternation of generations" (articles.directorym.com).

Algae are prokaryotic and eukaryotic. In prokaryotic cells, membrane-bounded organelles (nuclei, mitochondria, plastids, Golgi bodies and flagella) are not found and occur in the cyanobacteria (www.formatex.info). Eukaryotic algae have organelles (Tomaselli 2004; Lee 2008). Microalgae are able to sequester carbon dioxide from various sources in an efficient manner. Carbon dioxide fixation from atmosphere is one method to sequester carbon. It depends on the mass transfer from the air to algae during photosynthesis. But due to low amount of carbon dioxide in the atmosphere, the use of terrestrial plants is not an viable option (Brennan and Owende 2010; www.formatex.info). By contrast, industrial exhaust gases for example flue gas contain about 15% carbon dioxide, providing a carbon dioxide rich source for growing algae and a better route for biosequestration of carbon dioxide. Many algae can use sodium carbonate and sodium bicarbonate for growth (www.ijpdt.com). Some algae contain high extracellular carboanhydrase activities, which convert carbonate to free carbon dioxide for promoting carbon dioxide assimilation. In many algae, the uptake of bicarbonate by an active transport system has been observed (www.ijpdt.com; Wang et al. 2008). Inorganic elements which constitute the cell wall are provided by growth medium. Essential elements include nitrogen and phosphorus (www.formatex.info). "Using the molecular formula of the microalgal biomass ($CO0.48H1.83N0.11P0.01$), the minimal nutritional requirements can be found out", (www.formatex.info; Chisti 2007). Nitrogen is supplied as nitrate, ammonia and urea. Urea produces higher yield and does not cause much fluctuation in medium pH during the growth. Therefore, it is the preferred source (Shi et al. 2000; www.formatex.info).

Phosphates complex with metal ions so phosphorous is added in excess (Chisti 2007; www.ijpdt.com). Growth of algae depends on sufficient supply of important macronutrients, micronutrients and major ions (www.ijpdt.com; Sunda et al. 2005).

Autotrophic microalgae are able to convert the sun light into the carbon storage products, leading to lipids accumulation, including triacylglycerols, which are then converted into biofuels (Maity et al. 2014; eprints.usq.edu.au). Autotrophic microalgae is mostly grown in indoor photo-bioreactors and are used for production of biodiesel (Gordon and Seckbach 2012). However, autotrophic microalgae depends heavily on light for photosynthesis which result in higher energy outputs for illumination; a requirement for shallow cultivation systems with large surface areas (Geada et al. 2017; Kim et al. 2015, Mohan et al. 2015). In comparison there is a lot more flexibility in culturing of heterotrophic microalgae as they can grow without the addition of a light source and can store higher lipid contents in their cells (Zhang et al. 2013). In heterotrophic nutritional mode, organic molecules are used as carbon and energy source by microalgae; assist in high biomass yields and makes large-scale production much more feasible (ec.europa.eu). The relative simplicity of operations, easy maintenance and cost effectiveness are the main benefits during culturing of heterotrophic microalgae (Mohan et al. 2015).

The major producers in the aquatic ecosystem are microalgae (Becker 2004). They are single cell primitive organisms and have high capability to photosynthesize. They have a short growth period and get doubled in only 1–4 days. *Spirulina*

Table 3.1 Lipid contents of different microalgal species

Microalgae species	Lipid content (% dry weight)
Nannochloris sp.	20–56
Nannochloropsis sp.	12–53
Neochloris oleoabundans	29–65
Phaeodactylum tricornutum	18–57
Scenedesmus obliquus	11–55
Schizochytrium sp.	50–77
Botryococcus braunii	25–75
Chlorella	18–57
Chlorella emersonii	25–63
Chlorella sp.	10–48
Dunaliella sp.	18–67
Dunaliella tertiolecta	18–71

Based on Saifullah et al. (2014), Oncel (2013), Sakthivel et al. (2011)

Chlorella, and *Nitzschia* are generally used for production of biofuel (Chisti 2007; Algae for Biofuel Production 2014; Randhawa et al. 2016). Microalgae containing at least about 30% lipids can be used for production of biofuel. Microalgae are able to absorb carbon dioxide through photosynthesis and produce sugars and oxygen (www.diva-portal.org).

Photosynthesis can be shown as follows:

$$6CO_2 + 6H_2O + \text{sunlight energy} \rightarrow C_6H_{12}O_6 + 6O_2$$

Sugars can be converted to carbohydrates, proteins and lipids, which can be further converted to biofuel.

Algae grow quite rapidly and contain high content of oil compared with terrestrial crops. The terrestrial crops contain low oil content and take a season to grow (Chisti 2007; www.ijpdt.com). The doubling time of microalgae is 24 h. Some microalgae can double every three and one-half hours during the peak growth phase (Chisti 2007). Most of the algae which produce large amounts of lipids belong to Chorophyta, Cryptophyta, and Chromophyta divisions (Darzins et al. 2010; lrd. yahooapis.com) In microalgae, oil content varies between 20% and 50% whereas some strains contain upto 80% lipids (Spolaore et al. 2006; Metting 1996). Lipid contents of different microalgal species are presented in Table 3.1. Therefore, microalgae are attracting a lot of interest for production of biofuels (lrd.yahooapis.com).

References

Algae   for   Biofuel   Production   (2014)   articles.extension.org/pages/26600/
    algae-for-biofuel-production
Becker W (2004) Microalgae in human and animal nutrition. In: Handbook of microalgal culture.
    Blackwell, Oxford

Brennan L, Owende P (2010) Biofuels from microalgae – a review of technologies for production, processing, and extractions of biofuels and co-products. Renew Sust Energ Rev 14:557–577

Chisti Y (2007) Biodiesel from microalgae. Biotechnol Adv 25:294–306

Darzins A, Pienkos P, Edye L (2010) Current status and potential for algal biofuels production. T39-T2, IEA Bioenergy Task 39

Geada P, Vasconcelos V, Vicente A, Fernandes B (2017) Microalgal biomass cultivation. In: Rastogi R, Madamwar D, Pandey A (eds) Algal green chemistry: recent progress in biotechnology. Elsevier BV, Amsterdam, pp 257–284

Gordon R, Seckbach J (2012) The science of algal fuels. Springer, Dordrecht

Kim J, Lee JY, Lu T (2015) A model for autotrophic growth of Chlorella vulgaris under photolimitation and photoinhibition in cylindrical photobioreactor. Biochem Eng J 99:55–60

Lee RE (2008) Phycology, 4th edn. Cambridge University Press, Cambridge

Maity JP, Bundschuh J, Chen CY, Bhattacharya P (2014) Microalgae for third generation biofuel production, mitigation of greenhouse gas emissions and wastewater treatment: present and future perspectives – a mini review. Energy Elsevier 78(C):104–113

Metting FB (1996) Biodiversity and application of microalgae. J Ind Microbiol 17:477–489

Mohan SV, Rohit MV, Chiranjeevi P, Chandra R, Navaneeth B (2015) Heterotrophic microalgae cultivation to synergize biodiesel production with waste remediation: progress and perspectives. Bioresour Technol 184:169–178

Oncel SS (2013) Microalgae for a macroenergy world. Renew Sust Energ Rev 26:241–264

Randhawa KS, Relph LE, Armstrong MC, Rahman PKSM (2016) Biofuel production: tapping into microalgae despite challenges. Biofuels. Available online: 21 Sept 2016. https://doi.org/10.1080/17597269.2016.1224290

Saifullah AZA, Karim Md A, Ahmad-Yazid A (2014) Microalgae: an alternative source of renewable energy. Am J Eng Res 3(3):330–338

Sakthivel R, Elumalai S, Arif MM (2011) Microalgae lipid research, past, present: a critical review for biodiesel production, in the future. J Exp Sci 2(2011):29–49

Shi XM, Zhang XW, Chen F (2000) Heterotrophic production of biomass and lutein by Chlorella protothecoides on various nitrogen sources. Enzym Microb Technol 27:312–318

Spolaore P, Joannis-Cassan C, Duran E, Isambert A (2006) Commercial application of microalgae. J Biosci Bioeng 101:87–96

Sunda WG, Price NM, Morel FMM (2005) Trace metal ion buffers and their use in culture studies. In: Andersen RA (ed) Algal culturing techniques. Elsevier, Amsterdam, pp 35–63

Tomaselli L (2004) The microalgal cell. In: Richmond A (ed) Handbook of microalgal culture: biotechnology and applied phycology. Blackwell, Oxford, pp 3–19

Wang B, Li Y, Wu N, Lan C (2008) CO2 bio-mitigation using microalgae. Appl Microbiol Biotechnol 79:707–718

Zhang XL, Yan S, Tyagi RD, Surampalli RY (2013) Biodiesel production from heterotrophic microalgae through transesterification and nanotechnology application in the production. Renew Sust Energ Rev 26:216–223

aepc.gov.np
articles.directorym.com
biology.about.com
ec.europa.eu
eprints.usq.edu.au
lrd.yahooapis.com
oceanservice.noaa.gov
pubs.ext.vt.edu
www.dbbe.fcen.uba.ar
www.diva-portal.org
www.formatex.info
www.ijpdt.com
www.ret.gov.au
www.waterboards.ca.gov

Chapter 4
Cultivation of Third Generation Biofuel

Abstract Cultivation of microalgae in open ponds, closed ponds, photobioreactors and hybrid system are presented in this chapter. The most widely used photobioreactors – tubular, flat and column photobioreactors are discussed.

Keywords Cultivation · Microalgae · Open ponds · Closed ponds ·
Photobioreactors · Hybrid system · Biofuel · Circular ponds · Raceway ponds

Cultivation of microalgae is performed by using the open pond system and closed photo bioreactors system (www.oilgae.com; www.diva-portal.org): These systems have their own advantages and disadvantages. Some methods are able to produce a higher yield in comparison to other mrthods, and some methods can get easily contaminated (Xu et al. 2006; Show et al. 2017; Tüccar et al. 2015; Demirbas 2010a; Yoo et al. 2010; Li et al. 2007; Rodolfi et al. 2009; Jin et al. 2011; Molina Grima et al. 2001; Schenk et al. 2008; Tredici and Rodolfi 2004; Ugwu et al. 2008).

Microalgal culture system should possess the following characteristics from a commercial angle (www.mdpi.com; cest2015.gnest.org; Olaizola 2003):

- High area
- High productivity
- Less investment
- Lower maintenance costs
- Easier control of cultural parameters

Different designs of cultivation systems have been developed to obtain these characteristics (www.mdpi.com).

P. Bajpai, *Third Generation Biofuels*, SpringerBriefs in Energy,
https://doi.org/10.1007/978-981-13-2378-2_4

4.1 Open Ponds

The simplest systems for growing microalgae on a large scale are open ponds (Demirbas 2010a; Show et al. 2017; Tüccar et al. 2015; Fernandes et al. 2015). These systems have been used since the 1950s and widely examined in the past few years (Chaumont 1993; Tredici 2004; Terry and Raymond 1985; Spolaore et al. 2006; Borowitzka 1999).

The traditional type of open-air cultivation systems are presented below (www. formatex.info):

– Natural ponds and lakes
– Circular ponds
– Raceway ponds
– Inclined systems

The cheapest method for growing algae are open-air systems. These systems are widely used. These systems require less investment, the production capacity is larger and can operate longer than the closed systems. Most industrial systems currently used are of this type. They absorb sunlight and derive nutrients from runoff water from closeby land areas or from sewage treatment plants (Demirbas 2010a; Carlsson et al. 2007). These systems are extensively used at commercial level but present many challenges.

Open-air systems are operated in a continuous manner. In open-air systems, the shallow ponds are used which are designed in a raceway configuration. These ponds are generally 1 foot in depth. The growth conditions of algae are similar to their natural environment. For circulation and mixing of the nutrients with algal cells paddle wheels are used. The raceways are generally made up of concrete, or they are dug into the earth and lined with plastic to avoid the ground from absorbing the liquid. Baffles in the channel guide the flow around bends to reduce the space. The feed is mixed in front of the paddle wheel. Algae is harvested behind the paddle wheel after it circulates through the loop (lrd.yahooapis.com; Wen et al. 2011; articles.extension.org/pages/26600/algae-for-biofuel-production).

Depending on the nutrients required, several types of wastewater – municipal wastewater, dairy/swine lagoon effluent – can be used for the growing algae (lrd. yahooapis.com). Water with high salinity or seawater can be used for growing some marine microalgae. Raceway ponds are being used for growing *Arthrospira* by the major producers (Chini Zittelli et al. 2013).

Open ponds do not cost much to build and the performance is comparable to photobioreactors. But this system has many drawbacks. Due to evaporation, there is too much water loss because these are open-air systems (lrd.yahooapis.com). Therefore, microalgae do not use carbon dioxide efficiently when open ponds are used. There is lesser production of algal biomass in open ponds (Chisti 2007). Biomass productivity is less due to contamination with undesired algae and other microbes present in the feed. Furthermore maintaining the optimal culture conditions is difficult in open ponds. It is very expensive to recover the algae from the dilute algal broth (Molina Grima 1999; lrd.yahooapis.com). Only *Dunaliella* and *Spirulina* adaptable to very high salinity, and *Chlorella* adaptable to nutrient-rich media were cultivated in open pond on a commercial scale (Carlsson et al. 2007). Natural and artificial ponds are viable, when certain conditions are met,. The favorable environmental conditions and enough nutrients for microalgae to grow abundantly is unpreventable. It is also important that the water has high salinity, high concentration of nutrients, high pH, for avoiding the contamination (www.formatex.info).

Spirulina is being produced successfully in Lake Kossorom by open air cultivation method for food application (Abdulqader et al. 2000; www.formatex.info). About 40 ton/year is produced. Alkaline water is used in Myanmar for producing about 30 ton/year of *Spirulina* (Thein 1993). In Whyalla, South Australia, Betatene Ltd. is one of the important producers of *Dunaliella salina*. This company uses very large ponds up to 250 ha. There is no proper type of mixing other than by wind and convection (Chini Zittelli et al. 2013; Tredici 2004; Borowitzka 2005). A brackish and freshwater species of cyanobacteria *Aphanizomenon flos-aquae* are commercialized as nutraceuticals. These are harvested from the Upper Klamath Lake in Oregon, USA (Carmichael et al. 2000; Tredici et al. 2009).

In Japan, Taiwan and Indonesia circular ponds are used for cultivation of *Chlorella*. These ponds have a centrally pivoted rotating agitator. Sun Chlorella and Yaeyama *Chlorella* in Japan grow *Chlorella* in the ponds, whereas other companies – Chlorella Industry Co., Japan, use fermenters to produce the inoculum and then use the outdoor ponds for cultivation (Chini Zittelli et al. 2013). The circular design limits the pond size to about 10,000 m^2 due to the non homogeneous mixing and mechanical problems of a long rotating arm. The inclined system (cascades) concept was patented in 1999. It has been used widely and developed by the Institute of Microbiology of the Academy of Sciences of the Czech Republic (Chini Zittelli et al. 2013; Setlık et al. 1970; Doucha and Lıvansky 2009; riuma.uma.es).

In the inclined systems, the medium flows from the top to the bottom of a sloping surface and goes in a tank. Then it is pumped back and a high turbulence is produced. The optical path of the culture medium is short, so the light absorption is more effective and biomass concentration of 40–50 g/L are obtained and the harvesting becomes less expensive (Chini Zittelli et al. 2013; Masojıdek et al. 2011).

The outdoor open pond microalgae production systems are shown in Fig. 4.1.

4.2   Closed Ponds

The environmental conditions can be controlled in a better way in closed ponds as compared to the open ponds and many problems associated with an open pond are solved. There is more investment in case of closed ponds compared to the open ponds, but significantly less compared to photobioreactors. Closed ponds are generally used for growing *Spirulina* and are made using plexiglass. The idea behind the closed pond is to cover the pond with a greenhouse. It permits the species being grown to remain dominant and if unheated prolongs the growing season only slightly and if heated it can produce throughout the year. The amount of carbon

Fig. 4.1 Outdoor open pond microalgae production systems at scale. (**a**) Earthrise production facility, Calipatria, CA, and (**b–c**) Cyanotech facilities in Kailua Kona, HI. (IEA Bioenergy 2017. Reproduced with permission)

dioxide can be increased in these closed systems which increases the growth rate (http:www.oilgae.com research; ijcaonline.org; Tüccar et al. 2015).

4.3   Photobioreactors

Photobioreactors (PBR) are closed tank system. These reactors are used for growing phototrophs in which the photons are the major energy source for growth. The photons do not hit directly on the surface of culture but pass through the transparent

Table 4.1 Advantages of photobioreactors

Cultivation of algae is in controlled circumstances, hence potential for much higher productivity
Large surface-to-volume ratio. PBRs offer maximum efficiency in using light and therefore greatly improve productivity. Typically the culture density of algae produced is 10–20 times greater than bag culture in which algaeculture is done in bags – and can be even greater.
Better control of gas transfer.
Reduction in evaporation of growth medium.
More uniform temperature.
Better protection from outside contamination.
Space saving – Can be mounted vertically, horizontally or at an angle, indoors or outdoors.
Reduced Fouling – Recently available tube self cleaning mechanisms can dramatically reduce fouling.

Covering ponds does offer some of the benefits that are offered by photobioreactors, but enclosed systems will still provide better control of temperature, light intensity, better control of gas transfer, and larger surface area-to-volume ratio. An enclosed PBR design will enhance commercial algal biomass production by keeping algae genetics pure and reducing the possibility of parasite infestation.

Table 4.2 Disadvantages of photobioreactors

Capital cost is very high. This is one of the most important bottlenecks that is hindering the progress of algae fuel industry.
Despite higher biomass concentration and better control of culture parameters, data accumulated in the last two decades have shown that the productivity and production cost in some enclosed photobioreactor systems are not much better than those achievable in open-pond cultures.
The technical difficulty in sterilizing these photobioreactors has hindered their application for algae culture for specific end-products such as high value pharmaceutical products.

walls of the reactor before reaching the cells (Chini Zittelli et al. 2013; Dragone et al. 2010). In PBRs, the direct exchange of liquids, gases and particles between the culture and the atmosphere is strongly limited (Tredici et al. 2010). Compared with open-air systems, the PBR process is an attractive approach (Table 4.1) (Um and Kim 2009; Sydney et al. 2010; Scott et al. 2010; Molina Grima 1999; pubs.ext. vt.edu; mro.massey.ac.nz). Chisti (2007) has reported that on average, the productivity of PBR can be higher by 13 times compared to a traditional raceway pond. In PBR harvesting of biomass is cheaper in comparison to a raceway pond, because the biomass is more concentrated (about 30 times) (hd.yahooapis.com; Dudeja et al 2012; Chisti 2007). Enclosed PBR also have certain disadvantages (Table 4.2). These reactors are more expensive and scale up is difficult. Furthermore, it is not possible to overcome light limitation as the penetration of light is inversely proportional to the cell concentration. Penetration of light can be prevented by attaching the cells to the tube walls. High biomass concentration can be obtained in the closed system but the growth of microalgae is still suboptimal because of variations in light inresity and the temperature.

PBRs have been classified on the basis of (1) mode of operation and (2) design (jlakes.org). The reactors are flat or tubular; horizontal, inclined, vertical or spiral; manifold or serpentine; hybrid; floating and biofilm reactors (repositorio.uchile.cl). Tubular photobioreactor system is used for the indoor and outdoor cultivation (Liu et al. 2000; Amin 2009; www.diva-portal.org).

An operational classification of PBRs includes: air or pump-mixed and single-phase reactors or two-phase reactors (Chini Zittelli et al. 2013). In single-phase reactors; the exchange of gas takes place in a separate gas exchanger. In the two-phase reactors both liquid and gas are present and mass transfer of gas takes place in the reactor itself (docplayer.net).

Material of construction provide additional subcategories: Plastic or glass and rigid or flexible PBRs. Axenic reactors are used for operation under aseptic conditions and need to be sterilized before operation (Chini Zittelli et al. 2013).

Photobioreactors which are most widely used are discussed below:

4.3.1 Tubular Photobioreactor

Tubular PBRs are generally made-up of glass or plastic tube, The cultures are re-circulated either with pump or an airlift system. They can be in form of horizontal / serpentine, vertical, close to horizontal, conical, inclined photobioreactor (documents.mx). Tubular PBRs consist of straight, looped or coiled transparent tubing which are arranged in several ways for maximum absorption of light. Tubular photobioreactors designed properly completely isolate the culture from external contaminants and allow monoalgal culture for long duration (Basanta and Varma 2016).

Mixing and aeration in tubular PBRs are generally conducted by using airlift systems. These reactors are found suitable for outdoor cultivation of algae because they possess large surface area for illumination. But these bioreactors show poor mass transfer. This is one of their major limitations (documents.mx). When these reactors are scaled up, mass transfer becomes a problem (Soni et al. 2017). In tubular PBRs, it is possible to reach very high level of dissolved oxygen (www.orau.gov).

A common problem faced in case of outdoor tubular PBRs is photoinhibition. "When a tubular photobioreactor is scaled up by increasing the tube diameter, the illumination surface to volume ratio reduces. Conversely, the length of the tube can be kept shorter while a tubular PBR is scaled up by increasing the tube diameter. In this case, the cells present at the lower part of the tube will not get sufficient light for cell growth because of the light shading effect unless the mixing system is good. Also it is difficult to control temperature in most tubular PBRs. Thermostat can be used to maintain the desired temperature, but this could be difficult to implement and expensive" (documents.mx).

In tubular PBRs the cells are attached to the walls of the tubes. These reactors are characterized by gradients of carbon dioxide and oxygen transfer along the tubes. The increase in pH results in the re-carbonation of the cultures, increasing the production cost (www.formatex.info www.oilgae.com).

Most of the closed PBRs are tubular bioreactors. With *Spirulina*, using a 10 m³ serpentine bioreactor, productivity of 25 g m⁻² d⁻¹ was obtained with intermitted culture circulation (Torzillo et al. 1986 www.formatex.info; Dragone et al. 2010). Use of two-plane tubular PBR resulted in increased productivities of about 30 g m⁻² d⁻¹ (Torzillo et al. 1993). Helical tubular PBRs are a better option compared to straight tubular PBRs. Biocoil, is the commonly used reactor. It is marketed by Biotechna in Australia. This reactor contains polyethylene tubes (Carvalho et al. 2006). A 300 litre α-shaped tubular PBR for growing *Chlorella pyrenoidosa* was used by Lee et al. (1995). The system used an airlift pump for promoting an ascending/descending path, with many carbon dioxide injection points along its path. Transparent pipes made of glass or plastic are used in tubular reactor. The lighting conditions make the microalgae grow in the tubes which can be horizontally and vertically inclined (www.diva-portal.org). Similar to the open pond system, microalgae also need sunlight, carbon dioxide, water and some nutrients for growth. Natural sunlight can be used when the close PBRs are placed outside. But when these reactors are Open pond systems built inside, artificial lights are needed. Carbon dioxide should be fed into the PBRs as algae cannot absorb carbon dioxide from the air in the closed reactors,. Microalgae produces oxygen by photosynthesis. In the closed PBR oxygen cannot be released to the atmosphere as opposed to the open pond (www.diva-portal.org).It is very important that the system must remove excess oxygen from the reactors periodically (Chisti 2007; www.diva-portal.org).

4.3.2 Flat Photobioreactors

Flat-plate PBRs have large illumination surface area and are becoming popular for growing photosynthetic microorganisms (Milner 1953; Ugwu 2008; Samson and Leduy 1985; Ramos de Ortega and Roux 1986; documents.mx). Several designs of flat plate reactors and vertical alveolar panels for cultivating different algae on a commercial scale have been proposed (Tredici and Materassi 1992; Zhang et al. 2002; Hoekema et al. 2002; Soni et al. 2017). Generally, flat-plate PBRs are made-up of transparent materials for maximum absorption of light energy. Compared to horizontal tubular photobioreactors, dissolved oxygen concentrations is lower in flat-plate PBRs and photosynthetic efficiencies are higher (Hu et al. 1996; Richmond 2004;.documents.mx; www.ollgae.com, Soni et al. 2017; Ugwu 2008)

In the Flat PBRs, the dissolved oxygen is low and the photosynthetic efficiency is high in comparison to tubular designs. These reactors have been found suitable for growing microalgae on a large scale. In these reactors, a thin layer of dense culture is mixed across a flat transparent panel, allowing absorbance of radiation in the first few millimetres thickness (www.formatex.info; Brennan and Owende 2010; Mata et al. 2010; Geada et al. 2017; onlinelibrary.wiley.com). "The panels are generally illuminated mainly on one side by direct sunlight and they can be positioned vertically or inclined at an optimum angle facing the sun. This allows a better efficiency in terms of energy absorbed from incident sunlight. Packed flat panels mixed by air bubbling can potentially obtain very high overall productivities through lamination of solar light. There are certain limitations also which include problem in controlling temperature, possibility of hydrodynamic stress to some algal cultures, some degree of wall growth, scaling up requires support materials and many compartments" (www.formatex.info).

Transparent material is used for constructing flat-plated PBRs. The illumination surface area is large, so the efficiency of photosynthesis is high, the accumulation of dissolved oxygen is low, and the algae can be immobilized. Fabrication and maintenance of the reactors is easy. The large surface area presents difficulties in controlling temperature, problems in scale-up, and controlling diffusion rate of carbon dioxide and the tendency for attachment of the algae to the walls (Demirbas 2010a, b).

4.3.3   Column Photobioreactors

Several types of vertical-column PBRs (bubble-column and airlift photobioreactors) have been studied for growing algae. They appear to be very promising for producing algae on a large scale (documents.mx). These reactors are compact, low cost and can be operated monoseptically. Specific growth rate and final biomass concentration comparable to values achieved with narrow tubular PBRs can be obtained. Some type of bubble column PBRs are constructed as split cylinders or have draft tubes (documents.mx). Column PBRs show low energy consumption, reduced photo-oxidation and photoinhibition, high mass transfer, good mixing, easy to sterilize, easier control of growth conditions and immobilization of algae. Drawbacks are shear stress, small surface area for illumination; requirement of good materials for construction and reduction in illumination surface area when scaled-up. Their specific growth rate and final biomass concentration compares very well with the values obtained for tubular PBRs. Very high radial movement of fluid can be obtained in column reactors that is needed for better light–dark cycling. These reactor designs have a low surface/volume, but gas hold-ups are significantly greater compared to horizontal reactors. As a result, the effect of photo-inhibition and photo-oxidation on culture is less. The culture experience an adequate light–dark cycle (Mata et al. 2010; www.formatex.info; pdfs.semanticscholar.org; Ugwu 2008)

The closed PBR systems are shown in Fig. 4.2.

4.4   Hybrid Systems

Both open ponds and closed bioreactor system in combination are used for getting better results (Demirbas 2010a; Khan 2009). Open ponds are effective for growing algae, but they get contaminated very fast. A combination of both systems is the best method for growing algal strains for producing biofuels in a cost-effective manner

Fig. 4.2 Cultivation systems for algae growth. (**a–b**) flat panel photobioreactors at AzCATI),39 (**c**) Small, 1000 L research open ponds at AzCATI, and (**d e**) horizontal and vertical tubular photobioreactors at Algae PARC (between 12 and 24 m² each). (IEA Bioenergy 2017. Reproduced with permission)

(Demirbas 2010b). Open ponds are inoculated with a desired species grown in a bioreactor (www.bioaliment.ugal.ro; Demirbas 2010a). The inoculum size used should be large for the desired species to grow. Therefore, for reducing the contamination, cleaning of the ponds should be part of the aquaculture routine. Therefore open ponds can be considered as batch cultures (Demirbas 2010a, b).

References

Abdulqader G, Barsanti L, Tredici MR (2000) Harvest of Arthrospira platensis from Lake Kossorom (Chad) and its household usage among the Kanembu. J Appl Phycol 12:493–498

Amin S (2009) Review on biofuel oil and gas production processes from microalgae. Energy Convers Manag 50:1834–1840

Basanta KB, Varma A (2016) From algae to liquid fuels. In: Microbial resources for sustainable energy. Springer, Cham

Borowitzka MA (1999) Commercial production of microalgae: ponds, tanks, tubes and fermenters. J Biotechnol 70:313–321

Borowitzka MA (2005) Culturing microalgae in outdoor ponds. In: Andersen RA (ed) Algal culturing techniques. Elsevier Academic Press, Burlington, pp 205–218

Brennan L, Owende P (2010) Biofuels from microalgae – a review of technologies for production, processing, and extractions of biofuels and co-products. Renew Sust Energ Rev 14:557–577

Carlsson AS, van Beilen JB, Moller R, Clayton D (2007) Micro- and macro-algae: utility for industrial applications, 1st edn. CPL Press, Newbury

Carmichael WW, Drapeau C, Anderson DM (2000) Harvesting of Aphanizomenon flos-aquae Ralfs ex Born. & Flah. var. flos-aquae (Cyanobacteria) from Klamath Lake for human ietary use. J Appl Phycol 12:585–595

Carvalho AP, Meireles LA, Malcata FX (2006) Microalgal reactors: a review of enclosed system designs and performances. Biotechnol Prog 22:1490–1506

Chaumont D (1993) Biotechnology of algal biomass production: a review of systems for outdoor mass culture. J Appl Phycol 5:593–604

Chini Zittelli G, Biondi N, Rodolfi L, Tredici MR (2013) Photobioreactors for mass production of microalgae. In: Richmond A, Hu Q (eds) Handbook of microalgal culture: applied phycology and biotechnology, 2nd edn. Wiley, Oxford, pp 225–266

Chisti Y (2007) Biodiesel from microalgae. Biotechnol Adv 25:294–306

Demirbas A (2010a) Use of algae as biofuel sources. Energy Convers Manag 51:2738–2749

Demirbas A (2010b) Thermochemical processes. In: Biorefineries, Green energy and technology. Springer, London

Doucha J, Lıvansky K (2009) Outdoor open thin-layer microalgal photobioreactor: potential productivity. J Appl Phycol 21:111–117

Dragone G, Fernandes B, Vicente AA, Teixeira JA (2010) Third generation biofuels from microalgae. In: Mendez-Vilas A (ed) Current research, technology and education topics in applied microbiology and microbial biotechnology. Formatex, Badajoz, pp 1355–1366

Dudeja S, Bhattacherjee AB, Chela-Flores J (2012) Antarctica as model for the possible emergence of life on Europa. In: Hanslmeier A, Kempe S, Seckbach J (eds) Life on earth and other planetary bodies. Cellular origin and life in extreme habitats and astrobiology. Springer, Dordrecht

Fernandes BD, Mota A, Teixeira JA, Vicente AA (2015) Continuous cultivation of photosynthetic microorganisms: approaches, applications and future trends. Biotechnol Adv. https://doi.org/10.1016/j.biotechadv.2015.03.004

Geada P, Vasconcelos V, Vicente A, Fernandes B (2017) Microalgal biomass cultivation. Elsevier BV, Amsterdam

Hoekema S, Bijmans M, Janssen M, Tramper J, Wijffels RH (2002) A pneumatically agitated flatpanel photobioreactor with gas recirculation: anaerobic photoheterotrophic cultivation of a purple nonsulfur bacterium. Int J Hydrog Energy 27:1331–1338

Hu Q, Guterman H, Richmond A (1996) A flat inclined modular photobioreactor (FIMP) for outdoor mass cultivation of photoautotrophs. Biotechnol Bioeng 51:51–60

IEA Bioenergy (2017) State of technology review – algae bioenergy an IEA bioenergy inter-task strategic project. http://www.ieabioenergy.com/wp-content/uploads/2017/01/IEA-Bioenergy-Algae-report-update-20170114.pdf

Jin L, Huang J, Che F (2011) Microalgae as feedstocks for biodiesel production. In: Biodiesel – feedstocks and processing technologies. IntechOpen Limited, London

Khan SA (2009) Prospects of biodiesel production from microalgae in India. Renew Sust Energ Rev 13:2361–2372

Lee Y-K, Ding S-Y, Low C-S, Chang Y-C, Forday W, Chew P-C (1995) Design and performance of an α-type tubular photobioreactor for mass cultivation of microalgae. J Appl Phycol 7:47–51

Li X, Xu H, Wu Q (2007) Large-scale biodiesel production from microalga Chlorella protothecoides through heterotrophic cultivation in bioreactors. Biotechnol Bioeng 98:764–771

Liu ZW, Yu RQ, Guo Y (2000) Photobioreactors for cultivating microalgae. Modern Chem Ind 20(12):56–58

Masojɪdek J, Kopecky J, Giannelli L, Torzillo G (2011) Productivity correlated to photobiochemical performance of Chlorella mass cultures grown outdoors in thinlayer cascades. J Ind Microbiol Biotechnol 38:307–317

Mata TM, Martins AA, Caetano NS (2010) Microalgae for biodiesel production and other applications: a review. Renew Sust Energ Rev 14:217–232

Milner HW (1953) Rocking tray. In: Burlew JS (ed) Algal culture from laboratory to pilot plant, vol 600. Carnegie Institution, Washington, DC, p 108

Molina Grima E (1999) Microalgae, mass culture methods. In: Flickinger MC, Drew SW (eds) Encyclopedia of bioprocess technology: fermentation, biocatalysis and bioseparation. Wiley, New York, pp 1753–1769

Molina Grima E, Fern'andez J, Aci'en Fern'andez FG, Chisti Y (2001) Tubular photobioreactor design for algal cultures. J Biotechnol 92:113–131

Olaizola M (2003) Commercial development of microalgal biotechnology: from the test tube to the marketplace. Biomol Eng 20:459–466

Ramos De Ortega A, Roux JC (1986) Production of Chlorella biomass in different types of flat bioreactors in temperate zones. Biomass 10:141–156

Richmond A (2004) Biological principles of mass cultivation. In: Richmond A (ed) Handbook of microalgal cultures, biotechnology and applied phycology. Blackwell, Oxford, pp 125–177

Rodolfi L, Chini Zittelli G, Bassi N, Padovani G, Biondi N, Bonini G, Tredici MR (2009) Microalgae for oil: strain selection, induction of lipid synthesis and outdoor mass cultivation in a low-cost photobioreactor. Biotechnol Bioeng 102:100–112

Samson R, LeDuy A (1985) Multistage continuous cultivation of blue-green alga Spirulina maxima in the flat tank photobioreactors with recycle. Can J Chem Eng 63:105–112

Schenk PM, Thomas-Hall SR, Stephens E, Marx U, Mussgnug JH, Posten C (2008) Second generation biofuels: high-efficiency microalgae for biodiesel production. Bioenergy Res 1:20–43

Scott SA, Davey MP, Dennis JS, Horst I, Howe CJ, Lea-Smith DJ, Smith AG (2010) Biodiesel from algae: challenges and prospects. Curr Opin Biotechnol 21(3):277–286

Setlɪk I, Veladimir S, Malek I (1970) Dual purpose open circulation units for large scale culture of algae in temperate zones. I. Basic design considerations and scheme of pilot plant. Algol Stud 1:11

Show PL, Tang MSY, Nagarajan D, Ling TC, Ooi CW, Chang JS (2017) A holistic approach to managing microalgae for biofuel applications. Int J Mol Sci 18:215. https://doi.org/10.3390/ijms18010215

Soni RA, Sudhakar K, Rana RS (2017) Spirulina– from growth to nutritional product: a review. Trends Food Sci Technol 69:157–171

Spolaore P, Joannis-Cassan C, Duran E, Isambert A (2006) Commercial applications of microalgae. J Biosci Bioeng 101:87–96

Sydney EB, Sturm W, de Carvalho JC, Thomaz-Soccol V, Larroche C, Pandey A, Soccol CR (2010) Potential carbon dioxide fixation by industrially important microalgae. Bioresour Technol 101:5892–5896

Terry KL, Raymond LP (1985) System design for the autotrophic production of microalgae. Enzym Microb Technol 7:474–487

Thein M (1993) Production of Spirulina in Myanmar (Burma). Bulletin de l'Institut Océanographique 12:175–178

Torzillo G, Pushparaj B, Bocci F, Balloni W, Materassi R, Florenzano G (1986) Production of Spirulina biomass in closed photobioreactors. Biomass 11:61–74

Torzillo G, Carlozzi P, Pushparaj B, Montaini E, Materassi R (1993) A two-plane tubular photobioreactor for outdoor culture of Spirulina. Biotechnol Bioeng 42:891–898

Tredici MR (2004) Mass production of microalgae: photobioreactors. In: Richmond A (ed) Handbook of microalgal culture: biotechnology and applied phycology. Blackwell Science, Oxford, pp 178–214

Tredici MR, Materassi R (1992) From open ponds to vertical alveolar panels: the Italian experience in the development of reactors for the mass cultivation of photoautotrophic microorganisms. J Appl Phycol 4:221–231

Tredici MR, Rodolfi L (2004) Reactor for industrial culture of photosynthetic micro-organisms. PCT Patent WO2004/074423

Tredici MR, Biondi N, Chini Zittelli G, Ponis E, Rodolfi L (2009) Advances in microalgal culture for aquaculture feed and other uses. In: Burnell G, Allan G (eds) New technologies in aquaculture: improving production efficiency, quality and environmental management. Woodhead Publishing/CRC Press, Cambridge/Boca Raton, pp 610–676

Tredici MR, Chini Zittelli G, Rodolfi L (2010) Photobioreactors. In: Flickinger MC, Anderson S (eds) Encyclopedia of industrial biotechnology: bioprocess, bioseparation, and cell technology, vol 6. Wiley, Hoboken, pp 3821–3838

Tüccar G, Güngör C, Uludamar E, Aydin K (2015) The potential of microalgal biodiesel in Turkey. Energy Source Part B Econ Plann Policy 10(4):397–403

Ugwu CU (2008) Photobioreactors for mass cultivation of algae. Bioresour Technol 99:4021–4028

Ugwu CU, Aoyagi H, Uchiyama H (2008) Photobioreactors for mass cultivation of algae. Bioresour Technol 99:4021–4028

Um BH, Kim YS (2009) Review: a chance for Korea to advance algal-biodiesel technology. J Ind Eng Chem 15:1–7

Wen Z, Liu J, Chen F (2011) Biofuel from microalgae. In: Moo-Young M (ed) Comprehensive biotechnology. Elsevier BV, Amsterdam

Xu H, Miao X, Wu Q (2006) High quality biodiesel production from a microalga Chlorella protothecoides by heterotrophic growth in fermenters. J Biotechnol 126:499–507

Yoo C, Jun SY, Lee JY, Ahn CY, Oh HM (2010) Selection of microalgae for lipid production under high levels carbon dioxide. Bioresour Technol 101:71–74

Zhang K, Kurano N, Miyachi S (2002) Optimized aeration by carbon dioxide gas for microalgal production and mass transfer characterization in a vertical flat-plate photobioreactor. Bioproc Biosys Bioeng 25:97–101

cest2015.gnest.org
docplayer.net
documents.mx
jlakes.org
lrd.yahooapis.com
mro.massey.ac.nz
onlinelibrary.wiley.com
pdfs.semanticscholar.org
pubs.ext.vt.edu
repositorio.uchile.cl
research.ijcaonline.org
riuma.uma.es
www.bioaliment.ugal.ro
www.diva-portal.org
www.formatex.info
www.mdpi.com
www.oilgae.com
www.orau.gov
www.scribd.com/document/92252716/Micro-Algae

Chapter 5
Harvesting and Drying of Algal Biomass

Abstract Harvesting and drying of algal biomass is presented in this chapter. Bulk harvesting by flocculation, flotation and Sedimentation (Gravity settling) and concentration of algal slurry by Centrifugation, ultrasonic aggregation, filtration and electrophoresis are discussed.

Keywords Harvesting · Drying · Algal biomass · Bulk harvesting · Flocculation · Flotation · Sedimentation (gravity settling) · Centrifugation · Ultrasonic aggregation · Filtration · Electrophoresis

Harvesting of algae refers to concentration of diluted algal broth until a thick paste is obtained. It is one of the major challenges in *algae* biodiesel initiatives. According to Brennan and Owende (2010), this step accounts for about 20–30% of the overall production costs in algae-based biofuel production process. So, selection of a proper method for harvesting will have an impact on the overall economics of the process (www.mdpi.com; Behera and Varma 2016).

Harvesting of microalgae is difficult as the size of algae is small. Selection of the effective process depends on properties and size of algal strain (Behera and Varma 2016). Several physical and chemical methods can be used for harvesting (Wang et al. 2008; Demirbas 2010; Show et al. 2017; Shah et al. 2017; Behera and Varma 2016; www.formatex.info; mro.massey.ac.nz; Edzwald 1993; Muylaert et al. 2009; Pushparaj et al. 1993)

- Flocculation
- Centrifugation
- Filtration
- Ultrafiltration
- Air-flotation
- Autoflotation

The initial concentration of algae in the pond is in the range of 0.10–0.15% (v/v). After flocculation and settling, the concentration increases to 0.7%. Using a belt filter process, the concentration further increases to 2% (v/v). Drying of algae from 2% to 50% v/v requires about 60% of the energy content of the algae, which is very costly.

5.1   Bulk Harvesting

Bulk harvesting methods include flocculation, flotation and sedimentation by gravity and separate the biomass from the culture broth. The concentration factors are affected by the type of technology used and the initial concentration of biomass. These can be usually 100–800 times for reaching 2–7% of total solid content (Show et al. 2017; www.formatex.info; www.mdpi.com).

5.1.1   Flocculation

For initial dewatering, flocculation can be used that will substantially ease further processing. The cells aggregate and form flocs which settle very fast (Dragone et al. 2010). Harvesting of algae depends mainly on the size. Algae of smaller size can be settled and filtered more easily. Microalgal cells possess a negative charge. Due to this self-aggregation in suspension is prevented. The use of flocculants reduces or neutralises the negative surface charge. The flocculants coagulate the algae without having any effect on the composition and the quality (Molina Grima et al. 2003). Most commonly used metal salts are ferric chloride, aluminium sulphate and ferric sulphate (Brennan and Owende 2010; www.formatex.info. www.usea.org; Behera and Varma 2016).

The advantage in this method is chitosan which is a naturally occurring material can also be used as a bio-flocculant (Divakaran and Pillai 2002). Flocculation can be autoflocculation and chemical coagulation (Sukenik and Shelef 1984; Lee et al. 1998). The latter uses organic and inorganic coagulants (Uduman et al. 2010; Bilanovic et al. 1988). Combined flocculation has been also used (Sukenik et al. 1988). Some of the chemical flocculants however, can be expensive or toxic. This would increase the production cost or affect the product quality (Oh et al. 2001; www.mdpi.com).

5.1.2   Flotation

Flotation process was originally developed for the mineral industry (Show et al. 2017). Phoochinda and White (2003) reported the first application of the method for algae removal. The air gets converted into bubbles. The cells get attached to these bubbles which transport the cells to the water surface. No chemicals are required in this method and is more effective compared to the sedimentation process (Wang et al. 2008; Ndikubwimana et al. 2016). The small diameter paticles (less than 0.05 mm) can also be aggregated by this method (Yoon and Luttrell 1989). Also the cost for operating flotation is significantly less compared to centrifugation method. There are several types of flotation methods which have their own advantages and disadvantages (Table 5.1) (Nurdogan and Oswald 1996; Zimmerman et al. 2008; Ghernaout et al. 2015; Santander et al. 2011; Cheng et al. 2011; www.mdpi.com; mro.massey.ac.nz; Show et al. 2017):

DiAF is more energy efficient than DAF (Phoochinda and White 2003). Jet flotation shows about 98% of algal harvesting efficiency and reduces the resultant phosphorus content (Jameson 2004). Flotation has been considered as a potential method for harvesting but about its technical or economic viability not much evidence has been reported (Brennan and Owende 2010).

A method for froth flotation for harvesting of algae from dilute culture broth was developed by Levin et al. (1961). A long column is used for harvesting algae. The column contains the feed solution The column is aerated from below. Foam is generated which is harvested from a side arm which is placed near the top of the column. This method is not dependent upon the addition of floatants. The harvested cell concentration depends on the concentration of feed, rate of aeration, aerator porosity, pH and height of foam in the column. This process can be used for harvesting of algae on a large scale. The economics of this process is found to be favorable (www.science.gov; www.ncbi.nlm.nih.gov; arca.unive.it).

Wiley et al. (2009) reported that suspended air flotation (SAF) is better in comparison to DAF. Energy requirement and the ratio of air:solids is lower and loading rates are higher in SAF process. Use of SAF for separating the *Chlorella* and *Scenedesmus* biomass may reduce the cost of manufacturing of algal products such as fuel, fish food and fertilizer.

Liu et al. (1999) used DiAF process for harvesting *Chlorella* sp. "Two types of collector chemicals, cationic N-cetyl-N,N,N-trimethylammonium bromide (CTAB) and anionic sodium dodecylsulfate (SDS), were used. About 86% of the cells were removed with 40 mg/L of CTAB and 20% removal was obtained with 40 mg/L of SDS. Upon the addition of 10 mg/L of chitosan, more than 90% of the cell removal was achieved when SDS (20 mg/L) was used as the collector" (www.oilgae.com).

Flow rate affected the flotation slightly. pH values in the range of 4.0–5.0 were found to be optimum. When the ionic strength was high, flotation efficiency reduced. In the separation processes, the electrostatic interaction between the collector chemical and cell surface plays an important role.

Table 5.1 Flotation methods	
	Dissolved Air Flotation (DAF),
	Dispersed Air Flotation (Diaf)
	Electroflotation
	Jet Flotation
	Dispersed Ozone Flotation (Diof)

"*Algae Venture Systems*, LLC has developed a process for *harvesting*, *dewatering*, and *drying* (*HDD*) of *algae*. Algae Venture Systems harvester has two belts which move in opposite directions. Algal broth having the desired solid content is passed through a spout on to the top belt. This moves from left to right. The solid remains on top and water passes through the belt. A capillary belt moves in a countercurrent direction. This passes directly below the first belt. The capillary belt gets wet and draws the water through the top belt using liquid adhesion"(www.oilgae.com).

5.1.3   Sedimentation (Gravity Settling)

Sedimentation (Gravity settling) method uses the Stoke's law in its operation. This method is widely used for harvesting of algae particularly for treatment of wastewater, because of the reason that volume of wastewater used is large and biomass produced has a value (Chen et al. 1998; www.mdpi.com). The size of the cell and density effects the settling properties of the material aggregated (Schenk et al. 2008). Sometimes it is used along with flocculation for increasing the effectiveness of this process. It is found to be effective for *Spirulina* sp. having a large cell size (>70 µm). Cells having low density are not able to settle properly with this method (Munoz and Guieysse 2006; Wiley et al. 2009; www.mdpi.com).

5.2   Concentration

This method concentrates the slurry using centrifugation, filtration and ultrasonic aggregation methods (www.usea.org). This step consumes more energy than the bulk harvesting method (www.mdpi.com).

5.2.1   Centrifugation

This is a highly efficient method as its operation is very fast. It is energy-intensive. It depends on several factors (Heasman et al. 2000) such as settling properties of the cells and settling depth and the residence time of the slurry in the centrifuge. The recovery efficiency can be more than 95% (Heasman et al. 2000). There is also requirement for maintenance because of the free moving parts (Bosma et al. 2003). Furthermore, the particles get exposed to the high gravitational and shear forces. This may result in a substantial damage to the cells (Knuckey et al. 2006).

Centrifugal forces are applied in centrifugation for separating the biomass from the suspension. After separation, the microalgae can be removed by simply removing the excess medium (Harun et al. 2010; www.formatex.info). Centrifugation is an effective and fast method of recovering algal cells (Molina Grima et al. 2003; eprints.soton.ac.uk). But high shear and gravitational forces during centrifugation may harm the cells. This method is not economically viable because of high consumption of energy particularly when large volumes are used (www.formatex.info).

5.2.2   Ultrasonic Aggregation

Aggregation of the microalgal cells can be achieved using Ultrasonic sound. More than 90% separation and concentration factor of 20 times is achieved (Bosma et al. 2003). In the medical field, this method has been used with success (Otto et al. 2001; www.mdpi.com; www.researchgate.net).

5.2.3   Filtration

The most simple method for separation of large-size microalgal species is filtration (www.mdpi.com) and is found to be more competitive compared to other harvesting methods. Pressure pump and a filter sheet are required for filtering the microalgae. Filtration methods are an attractive dewatering option but the running costs are high. Different methods have been used for filtration (Table 5.2).

Filtration involves passing the algal broth through the filters on which the algae gets accumulated and the medium passes through the filter. The broth is continuously passed through the filters until a thick algal paste is obtained (www.formatex. info). With this method, a concentration factor of 245 times was achieved using *Coelastrum proboscideum* (Mohn 1980).

Membranes have been also used for harvesting algae. For smaller size algae, the membrane needs to be changed (Petrusevski et al. 1995).The membrane filters are costly. Therefore, in comparison to the centrifugation method this method is less economical (MacKay and Salusbury 1988). The filtration membranes can also get contaminated which affects the product quality. Surface charge of cells cause polarization phenomenon. This may also change the cell characteristics, or the surface of the membrane (Phoochinda and White 2003).

Table 5.2 Methods for filtration

Dead end filtration
Microfiltration
Ultra filtration
Pressure filtration
Vacuum filtration
Tangential flow filtration

5.2.4   *Electrophoresis*

This method is safe, cost-effective and efficient (Show et al. 2017). Based on negative charge, which occurs naturally, the cells are separated by applying an electric
field in the electrophoresis process (Mollah et al. 2004). The method can be adjusted
easily by increasing the electrical power to speed up the segregation process
(Alfafara et al. 2002; www.mdpi.com; Show et al. 2017).

References

Alfafara CG, Nakano K, Nomura N, Igarashi T, Matsumura M (2002) Operating and scale-up
    factors for the electrolytic removal of algae from eutrophied lakewater. J Chem Technol
    Biotechnol 77:871–876
Behera BK, Varma A (2016). From algae to liquid fuels. In: Microbial resources for sustainable
    energy. Springer, Cham
Bilanovic D, Shelef G, Sukenik A (1988) Flocculation of microalgae with cationic polymers –
    effects of medium salinity. Biomass 17:65–76
Bosma R, van Spronsen WA, Tramper J, Wijffels RH (2003) Ultrasound,.a new separation tech-
    nique to harvest microalgae. J Appl Phycol 15:143–153
Brennan L, Owende P (2010) Biofuels from microalgae – a review of technologies for production,
    processing, and extractions of biofuels and co-products. Renew Sust Energ Rev 14:557–577
Chen Y, Liu J, Ju YH (1998) Flotation removal of algae from water. Colloids Surf B Biointerfaces
    12:49–55
Cheng YL, Juang YC, Liao GY, Tsai PW, Ho SH, Yeh KL, Chen CY, Chang JS, Liu JC, Chen WM
    (2011) Harvesting of Scenedesmus obliquus FSP-3 using dispersed ozone flotation. Bioresour
    Technol 102:82–87
Demirbas A (2010) Use of algae as biofuel sources. Energy Convers Manag 51:2738–2749
Divakaran R, Pillai VS (2002) Flocculation of algae using chitosan. J Appl Phycol 2002(14):419–422
Dragone G, Fernandes B, Vicente AA, Teixeira JA (2010) Third generation biofuels from micro-
    algae. In: Mendez-Vilas A (ed) Current research, technology and education topics in applied
    microbiology and microbial biotechnology. Formatex, Badajoz, pp 1355–1366
Edzwald J (1993) Algae, bubbles, coagulants, and dissolved air flotation. Water Sci Technol
    27:67–81
Ghernaout D, Benblidia C, Khemici F (2015) Microalgae removal from Ghrib Dam (Ain Defla,
    Algeria) water by electroflotation using stainless steel electrodes. Desalination Water Treat
    2015(54):3328–3337
Grima EM, Belarbi EH, Fernández FA, Medina AR, Chisti Y (2003) Recovery of microalgal bio-
    mass and metabolites: process options and economics. Biotechnol Adv 20:491–515
Harun R, Singh M, Forde GM, Danquah MK (2010) Bioprocess engineering of microalgae to
    produce a variety of consumer products. Renew Sust Energ Rev 14:1037–1047
Heasman M, Diemar J, O'connor W, Sushames T, Foulkes L (2000) Development of extended
    shelf-life microalgae concentrate diets harvested by centrifugation for bivalve molluscs – a
    summary. Aquac Res 2000(31):637–659
Jameson GJ (2004) Application of the Jameson cell technology for algae and phosphorus removal
    from maturation ponds. Int J Miner Process 73:23–28
Knuckey RM, Brown MR, Robert R, Frampton DM (2006) Production of microalgal concentrates
    by flocculation and their assessment as aquaculture feeds. Aquac Eng 35:300–313
Lee S, Kim S, Kim J, Kwon G, Yoon B, Oh H (1998) Effects of harvesting method and growth
    stage on the flocculation of the green alga Botryococcus braunii. Lett Appl Microbiol 27:14–18

Levin GV, Clendenning JR, Gibor A, Bogar FD (1961) Harvesting of algae by froth flotation. Appl Environ Microbiol 10(2):169–175

Liu J, Chen Y, Ju Y (1999) Separation of algal cells from water by column flotation. Sep Sci Technol 34:2259–2272

MacKay D, Salusbury T (1988) Choosing between centrifugation and crossflow microfiltration. Chem Eng 447:45–50

Mohn F (1980) Experiences and strategies in the recovery of biomass from mass cultures of microalgae. Algae biomass: production and use/sponsored by the National Council for Research and Development, Israel and the Gesellschaft fur Strahlen-und Umweltforschung (GSF), Munich, Germany; Shelef G, Soeder CJ (Eds) Food and Agriculture Organization of the United Nations: Rome, Italy

Mollah MY, Morkovsky P, Gomes JA, Kesmez M, Parga J, Cocke DL (2004) Fundamentals, present and future perspectives of electrocoagulation. J Hazard Mater 2004(114):199–210

Munoz R, Guieysse B (2006) Algal-bacterial processes for the treatment of hazardous contaminants: a review. Water Res 40:2799–2815

Muylaert K, Vandamme D, Meesschaert B, Foubert I (2009) Flocculation of microalgae using cationic starch. In: Phycologia. The Physiological Society, London, p 63

Ndikubwimana T, Chang J, Xiao Z, Shao W, Zeng X, Ng IS, Lu Y (2016) Flotation: a promising microalgae harvesting and dewatering technology for biofuels production. Biotechnol J 11:315–326

Nurdogan Y, Oswald WJ (1996) Tube settling of high-rate pond algae. Water Sci Technol 33:229–241

Oh HM, Lee SJ, Park MH, Kim HS, Kim HC, Yoon JH, Kwon GS, Yoon BD (2001) Harvesting of Chlorella vulgaris using a bioflocculant from Paenibacillus sp. AM49. Biotechnol Lett 23:1229–1234

Otto C, Baumann M, Schreiner T, Bartsch G, Borberg H, Schwandt P, Schmid-Schönbein H (2001) Standardized ultrasound as a new method to induce platelet aggregation: evaluation, influence of lipoproteins and of glycoprotein IIb/IIIa antagonist tirofiban. Eur J Ultrasound 14:157–166

Petrusevski B, Bolier G, van Breemen A, Alaerts G (1995) Tangential flow filtration: a method to concentrate freshwater algae. Water Res 29:1419–1424

Phoochinda W, White DA (2003) Removal of algae using froth flotation. Environ Technol 2003(24):87–96

Pushparaj B, Pelosi E, Torzillo G, Materassi R (1993) Microbial biomass recovery using a synthetic cationic polymer. Bioresour Technol 43:59–62

Santander M, Rodrigues RT, Rubio J (2011) Modified jet flotation in oil (petroleum) emulsion/water separations. Colloids Surf A Physicochem Eng Asp 375:237–244

Schenk M, Thomas-Hall SR, Stephens E, Marx UC, Mussgnug JH, Posten C, Kruse O, Hankamer B (2008) Second generation biofuels: high-efficiency microalgae for biodiesel production. Bioenergy Res 1:20–43

Shah JH, Deokar A, Patel K, Panchal K, Alpesh V, Mehta AV (2017) A comprehensive overview on various method of harvesting microalgae according to Indian. Perspective, international conference on multidisciplinary research & practice. Volume I, Issue VII, p 313. IJRSI. ISSN 2321-2705

Show PL, Tang MSY, Nagarajan D, Ling TC, Ooi CW, Chang JS (2017) A holistic approach to managing microalgae for biofuel applications. Int J Mol Sci 18:215. https://doi.org/10.3390/ijms18010215

Sukenik A, Shelef G (1984) Algal autoflocculation – verification and proposed mechanism. Biotechnol Bioeng 26:142–147

Sukenik A, Bilanovic D, Shelef G (1988) Flocculation of microalgae in brackish and sea waters. Biomass 15:187–199

Uduman N, Qi Y, Danquah MK, Forde GM, Hoadley A (2010) Dewatering of microalgal cultures: a major bottleneck to algae-based fuels. J Renew Sustain Energy 2:12701

Wang B, Li Y, Wu N, Lan CQ (2008) CO$_2$ bio-mitigation using microalgae. Appl Microbiol Biotechnol 79:707–718

Wiley P, Brenneman K, Jacobson A (2009) Improved algal harvesting using suspended air flotation. Water Environ Res 81(7):702–708

Yoon R, Luttrell G (1989) The effect of bubble size on fine particle flotation. Miner Process Extr Metall Rev 5:101–122

Zimmerman WB, Tesar V, Butler S, Bandulasena HC (2008) Microbubble generation. Recent Pat Eng 2:1–8

arca.unive.it

eprints.soton.ac.uk

mro.massey.ac.nz

www.e-education.psu.edu

www.formatex.info

www.mdpi.com

www.ncbi.nlm.nih.gov

www.oilgae.com

www.researchgate.net

www.science.gov

www.usea.org

Chapter 6
Extraction of Oil from Algal Biomass

Abstract Extraction of oil from algal biomass is presented in this chapter. Expeller/ oil pressing; Single solvent extraction; Supercritical carbon dioxide extraction; Microwave-assisted extraction; Enzymatic extraction; Ultrasound assisted extraction; Oxidative stress; Electroporation; Osmotic shock are discussed.

Keywords Oil extraction · Algal biomass · Expeller/oil pressing · Single solvent extraction · Supercritical carbon dioxide extraction · Microwave-assisted extraction · Enzymatic extraction · Ultrasound assisted extraction · Oxidative stress · Electroporation · Osmotic shock

Extraction is an important step to the use of algae for producing fuels (www.e-education.psu.edu). For release of the desired products, the cells are broken down. The triacylglycerols are present in oil droplets within the algal cells, and the cell wall presents a significant barrier to oil removal (www.rsisinternational.org). The steps involved in the extraction of oil from the cells are rupturing the cell wall and separating the oil from the biomass. Then the oil is purified or upgraded to remove the impurities. Table 6.1 show the different methods used for extraction of oil.

Naghdi et al. (2016) have published an excellent article on various techniques for extraction of lipids from algal biomass. The common methods used are – solvent extraction, supercritical fluid extraction, and ultrasonic assisted extraction. The oil yield from the biomass depends on the efficacy of the method used for extraction (Ryan 2009; www.formatex.info).

6.1 Expeller/Oil Pressing

This is a mechanical process for extraction of oil from raw materials such as nuts and seeds. High pressure presses are used for squeezing and breaking the cells. This process can be made effective by drying the algae first. This method does not require

Table 6.1 Methods used for extraction of oil from algae

Physical methods
1) Mechanical disruption (i.e., bead mills)
2) Electric fields
3) Sonication
4) Osmotic shock
5) Expeller press
Chemical and biological methods
1) Solvent extraction using single solvent, co-solvent, and direct reaction by transesterification
2) Supercritical fluids
3) Enzymatic extraction

Based on www.e-education.psu.edu

any special skills and can recover 75% of oil. It is found to be less effective because the extraction time is long (www.formatex.info; Harun et al. 2010).

6.2 Single Solvent Extraction

This is the common method and commercially used. Extraction is performed at high temperatures and pressure. Hexane or petroleum are used as solvents. These solvents are chemically similar to the lipids (www.e-education.psu.edu). Solvent raptures the cell wall, and extract oil from the medium as they are highly soluble in organic solvents in comparison to water. Distillation process is used for separating the oil from solvent. The solvent can be recovered and used further (www.formatex. info). The mass transfer rate and solvent accessibility increases and the dielectric constant of immiscible solvent reduces (www.e-education.psu.edu). The co-solvent process differs slightly. For selection of a co-solvent, criteria used are to use a more polar co-solvent that breaks the cell wall, and to use a second less polar co-solvent for matching the polarity of the lipids being extracted. This criteria is being met by alkanes (www.e-education.psu.edu). Several examples of co-solvent extraction are available. Bligh and Dyer developed a method in 1959. Solvents used were alcohol and chloroform. Most of the lipids dissolve into the chloroform. The interactions include water/methanol > methanol/chloroform > lipid/chloroform. Other combinations of co-solvents include (www.e-education.psu.edu):

– Dimethyl sulfoxide/petroleum ether
– Hexane/isopropanol
– Hexane/ethanol

Mixing of the polar and non-polar solvents increase polarity of the solvent which can improve the lipid extraction/recovery (Naghdi et al. 2016). Polar solvents actually help in releasing the lipids from protein–lipid complexes which helps their dissolving in the non-polar solvent (Ryckebosch et al. 2012). By performing the lipid extraction/recovery with wet biomass, this effect can be increased. When the wet biomass is used, the polar solvent is able to enter the water layer and makes the lipids available for salvation with non-polar solvent (Yoo et al. 2012; Naghdi et al. 2014). The recovery of total lipids and the total fatty acid methyl ester can be increased with a mixture of polar and non-polar solvents (Naghdi et al. 2016). Use of a solvent mixture of hexane and ethanol at a ratio of 3:1 respectively, significantly increased fatty ester methyl esters recovery by 50% in case of *Tetraselmis* sp. (Naghdi et al. 2016). The cost of Hexane is low and it shows the highest extraction capability (Harun et al. 2010; Potumarthi and Baadhe 2013).

6.3 Supercritical Carbon Dioxide Extraction (SCCO$_2$)

This is a green technology and can replace organic solvents used for extraction of lipids (Naghdi et al. 2016). This technology can also extract triacylglycerides and other lipids. Solvent-free extract is produced in less time showing reduced toxicity in comparison with the organic solvents (Andrich et al. 2005; Soh and Zimmerman 2011). SCCO$_2$ extraction with hexane extraction of lipids from *Chlorococcum* sp. was examined (Halim et al. 2011). Extraction with hexane was found to be less efficient as it required a very long time for obtaining yield comparable to SCCO$_2$. This technique is however, costly due to its high energy consumption, required infrastructure and operation (Naghdi et al. 2016).

6.4 Microwave-Assisted Extraction

This technique was developed by Ganzler et al. (1986) in the mid 1980s for obtaining lipids and pesticides from foods, feeds, seeds and soil (journal.frontiersin.org). For microalgal cultures, this technology is safe and rapid and reduces the costs of water removal and extraction of dry biomass (Naghdi et al. 2016; Amarni and Kadi 2010; Refaat et al. 2008). Naghdi et al. (2016) reports that, "the contact between a dielectric or polar material and a rapidly oscillating electric field (produced by microwaves) produces heat due to frictional forces which arise from inter- and intra-molecular movements. As heat is produced, water vapour begins to form within the cell, eventually rupturing it, leading to an electroporation effect which further opens up the cell membrane and releases the intracellular contents". In *Chlorella vulgaris*, use of microwaves along with sonication, produced higher lipid yield (Šoštarič et al. 2012). "Microwave irradiation also helps in the transesterification process post extraction by substituting conventional heating". Several factors should be taken into consideration when implementing the microwave technique (Naghdi et al. 2016;

Eskilsson and Björklund 2000). These are temperature, time, ratio of solid and liquid, type and concentration of the solvent and dielectric properties of the mixture.

This method is found to be effective for extraction of wet lipids and the extraction of lipids of high quality. But maintenance cost is a limiting factor (Naghdi et al. 2016).

6.5   Enzymatic Extraction

This technique is similar to solvent extraction. An enzyme is used for separating the materials instead of a solvent (www.e-education.psu.edu). This method results in a good lipid recovery (Naghdi et al. 2016). As the enzymes are selective, the cells are disrupted with the minimum damage to the target product (Demuez et al. 2015). Fu et al. (2010) studied immobilized cellulase for breaking *Chlorella* sp. The extraction efficiency of lipids was 56% which was higher by 14% in comparison with unhydrolysed microalgae. Similarly, use of cellulase enzyme on *C. vulgaris* cultures increased lipid extraction by 1.73 times in comparison with unhydrolysed cultures (Cho et al. 2013). Treatment of *C. vulgaris* with snailase, lysozyme and cellulase resulted in lipid recovery of 7%, 22% and 24% respectively (Zheng et al. 2011). With lysozyme, highest yield of 16.6% was obtained (Taher et al. 2014). With snailase and trypsin enzymes, lipid recovery was 35% whereas with cellulase, neutral protease and alkaline protease enzymes, lipid recovery was 16%, 12% and 8% respectively (Liang et al. 2012).

6.6   Ultrasound Assisted Extraction

Application of ultrasounds has also been examined for oil extraction. This approach seems to have a great potential (Harun et al. 2010; Gharabaghi et al. 2015; Metherel et al. 2009; Vinatoru et al. 1997; Ranjith Kumar et al. 2015; Chemat et al. 2004). Algae is exposed to ultrasonic waves of high intensity which generates small cavitation bubbles around cells. The bubbles are collapsed, the shockwaves are emitted, breaking the cell wall and the desired products are released (www.formatex.info). This method is used extensively at laboratory scale but not much information on cost aspects are available.

"Increase in exposure time can result in higher lipid yield, which is increased further using a mixture of polar and non-polar solvents. This technique not only reduces extraction time but also promotes the absorption of cell contents into the solvent through mass transfer and penetration of the solvent into the cell" (Naghdi et al. 2016). Use of ethanol improves environmental safety. Ethanol is better than MAE as it can be used at low temperatures; the thermal denaturation of biomolecules is reduced. When the microalgal cultures are subjected to sonication for a long time, free radicals are generated which harm the quality of the lipids. The non-polar organic solvents which are not susceptible to peroxide formation limit the oxidation

(Naghdi et al. 2016). The ratio of extraction solvents for ultrasound extraction is 2:1 chloroform: methanol and 3:2 hexane: isopropanol in flaxseed. This reduces oxidation of lipids and results in higher yields (Metherel et al. 2009).

6.7   Oxidative Stress

Lipid extraction efficiency can be improved by UV light or oxidative agents (Sharma et al. 2014). Oxidative stress using free nitrous acid (FNA) as pre-treatment for oil extraction was examined by Bai et al. (2014). A lipid yield higher by 2.4-times for algae treated with up to 2.19 mg HNO_2-N/L was obtained. This approach appears to be promising. Further research is needed. FNA is considered an environment friendly chemical (Wang 2013).

6.8   Electroporation

Algal cells are exposed to an electrical field for increasing the cell membrane permeability allowing chemicals, drugs, or DNA to be introduced into the cell (www.omicsgroup.org; Ho and Mittal 1996; Naghdi et al. 2016). Not many studies have been conducted in this area. Total lipid extraction of 92% from *Pseudochlorococcum* sp. Was achieved as compared to 62% with the Bligh and Dyer method (Sommerfeld et al. 2008).

6.9   Osmotic Shock

Several researchers have studied the lipid extraction after osmotic shock (Lee et al. 2010; Prabakaran and Ravindran 2011; Yoo et al. 2012; Mercer and Armenta 2011). When the osmotic shock is applied, algae cells rapture and release their contents because of substantial reduction of osmotic pressure (Naghdi et al. 2016). This method depends on cell wall properties, and it is not widely used. Recovery of lipids from *Chlamydomonas reinhardtii* by osmotic shock using organic solvents was examined by Yoo et al. (2012). The recovery of lipids was increased by about two times by osmotic shock.

References

Amarni F, Kadi H (2010) Kinetics study of microwave-assisted solvent extraction of oil from olive cake using hexane: comparison with the conventional extraction. Innov Food Sci Emerg Technol 11:322–327

Andrich G, Nesti U, Venturi F, Zinnai A, Fiorentini R (2005) Supercritical fluid extraction of bioactive lipids from the microalga *Nannochloropsis* sp. Eur J Lipid Sci Technol 107:381–386

Bai X, Naghdi FG, Ye L, Lant P, Pratt S (2014) Enhanced lipid extraction from algae using free nitrous acid pretreatment. Bioresour Technol 159:36–40

Bligh EG, Dyer WJ (1959) A rapid method of total lipid extraction and purification. Can J Biochem Physiol 37(8):911–917

Chemat F, Grondin I, Costes P, Moutoussamy L, Sing ASC, Smadja J (2004) High power ultrasound effects on lipid oxidation of refined sunflower oil. Ultrason Sonochem 11:281–285

Cho HS, Oh YK, Park SC, Lee JW, Park JY (2013) Effects of enzymatic hydrolysis on lipid extraction from *Chlorella vulgaris*. Renew Energy 54:156–160

Demuez M, Mahdy A, Tomás-Pejó E, González-Fernández C, Ballesteros M (2015) Enzymatic cell disruption of microalgae biomass in biorefinery processes. Biotechnol Bioeng 112:1955–1966

Eskilsson CS, Björklund E (2000) Analytical-scale microwave-assisted extraction. J Chromatogr A 902:227–250

Fu CC, Hung TC, Chen JY, Su CH, Wu WT (2010) Hydrolysis of microalgae cell walls for production of reducing sugar and lipid extraction. Bioresour Technol 101:8750–8754

Ganzler K, Salgo A, Valkó K (1986) Microwave extraction: a novel sample preparation method for chromatography. J Chromatogr A 371:299–306

Gharabaghi M, Amrei HD, Zenooz AM, Guzullo JS, Ashtiani FZ (2015) Biofuels: bioethanol, biodiesel, biogas, biohydrogen from plants and microalgae. In: Environmental chemistry for a sustainable world volume 5. Springer, Dordrecht, pp 233–274

Halim R, Gladman B, Danquah MK, Webley PA (2011) Oil extraction from microalgae for biodiesel production. Bioresour Technol 102:178–185

Harun R, Singh M, Forde GM, Danquah MK (2010) Bioprocess engineering of microalgae to produce a variety of consumer products. Renew Sust Energ Rev 14:1037–1047

Ho S, Mittal G (1996) Electroporation of cell membranes: a review. Crit Rev Biotechnol 16:349–362

Lee JY, Yoo C, Jun SY, Ahn CY, Oh HM (2010) Comparison of several methods for effective lipid extraction from microalgae. Bioresour Technol 101:S75–S77

Liang K, Zhang Q, Cong W (2012) Enzyme-assisted aqueous extraction of lipid from microalgae. J Agric Food Chem 60:11771–11776

Mercer P, Armenta RE (2011) Developments in oil extraction from microalgae. Eur J Lipid Sci Technol 113:539–547

Metherel AH, Taha AY, Izadi H, Stark KD (2009) The application of ultrasound energy to increase lipid extraction throughput of solid matrix samples (flaxseed). Prostaglandins Leukot Essent Fat Acids 81:417–423

Naghdi GF, Thomas-Hall SR, Durairatnam R, Pratt S, Schenk PM (2014) Comparative effects of biomass pre-treatments for direct and indirect transesterification to enhance microalgal lipid recovery. Front Energy Res 2:57

Naghdi FG, Gonzalez LMG, Chan W, Schenk PM (2016) Progress on lipid extraction from wet algal biomass for biodiesel production. Microb Biotechnol 9(6):718–726

Potumarthi R, Baadhe R (2013) Issues in algal biofuels for fuelling the future. In: In book: applications of microbial engineering. CRC Press, Boca Raton, pp 408–425

Prabakaran P, Ravindran A (2011) A comparative study on effective cell disruption methods for lipid extraction from microalgae. Lett Appl Microbiol 53:150–154

Ranjith Kumar R, Hanumantha Rao P, Arumugam M (2015) Lipid extraction methods from microalgae: a comprehensive review. Front Energy Res 2:61

Refaat A, El Sheltawy S, Sadek K (2008) Optimum reaction time, performance and exhaust emissions of biodiesel produced by microwave irradiation. Int J Environ Sci Technol 5:315–322

Ryan C (2009) Cultivating clean energy: the promise of algae biofuels. Report to Natural Resources Defense Council, USA, October 2009

Ryckebosch E, Muylaert K, Foubert I (2012) Optimization of an analytical procedure for extraction of lipids from microalgae. J Amer Oil Chem Soc 89:189–198

Sharma K, Li Y, Schenk PM (2014) UV-C-mediated lipid induction and settling, a step change towards economical microalgal biodiesel production. Green Chem 16:3539–3548

Soh L, Zimmerman J (2011) Biodiesel production: the potential of algal lipids extracted with supercritical carbon dioxide. Green Chem 13:1422–1429

Sommerfeld M, Chen W, Hu Q, Giorgi D, Navapanich T, Ingram M, Erdman R (2008) Application of electroporation for lipid extraction from microlalgae. Algae Biomass Summit, Seattle

Šoštarič M, Klinar D, Bricelj M, Golob J, Berovič M, Likozar B (2012) Growth, lipid extraction and thermal degradation of the microalga *Chlorella vulgaris*. New Biotechnol 29:325–331

Taher H, Al-Zuhair S, Al-Marzouqi AH, Haik Y, Farid M (2014) Effective extraction of microalgae lipids from wet biomass for biodiesel production. Biomass Bioenergy 66:159–167

Vinatoru M, Toma M, Radu O, Filip P, Lazurca D, Mason T (1997) The use of ultrasound for the extraction of bioactive principles from plant materials. Ultrason Sonochem 4(2):135–139

Wang Y (2013) Microalgae as the third generation biofuel: production, usage, challenges and prospects. Uppsala University, Uppsala

Yoo G, Park WK, Kim CW, Choi YE, Yang JW (2012) Direct lipid extraction from wet *Chlamydomonas reinhardtii* biomass using osmotic shock. Bioresour Technol 123:717–722

Zheng H, Yin J, Gao Z, Huang H, Ji X, Dou C (2011) Disruption of *Chlorella vulgaris* cells for the release of biodiesel-producing lipids: a comparison of grinding, ultrasonication, bead milling, enzymatic lysis, and microwaves. Appl Biochem Biotechnol 164:1215–1224

journal.frontiersin.org

www.e-education.psu.edu

www.formatex.info

www.omicsgroup.org

www.rsisinternational.org

Chapter 7
Production of Biofuel from Microalgae

Abstract Production of biofuel from microalgae is presented in this chapter. Biochemical Conversion (Anaerobic Digestion; Alcoholic Fermentation; Hydrogen production; Biodiesel production) and thermochemical conversion (Gasification; Hydrothermal liquefaction; Hydrothermal carbonization; Pyrolysis) processes are discussed.

Keywords Biofuel · Microalgae · Biochemical conversion · Anaerobic digestion · Alcoholic fermentation · Hydrogen production · Biodiesel production · Thermochemical conversion · Gasification · Hydrothermal liquefaction · Hydrothermal carbonization · Pyrolysis

Biofuels can be produced from microalgae in an economically effective and environmentally sustainable way. Algae can be used to produce third generation biofuel. The yield of algal fuels is higher in comparison to second generation biofuels. Compared to terrestrial crops, algae can produce 30–100 times more energy per hectare. Different types of biofuels have been produced from microalgae (Demirbas 2010d) (Table 7.1). Microalgae commonly used for production of biofuel are listed in Table 7.2.

Biofuel production can be coupled with wastewater treatment, production of value added products and flue gas carbon dioxide alleviation. New developments in growing microalgae and downstream processing would further increase the cost effectiveness of algal biofuel production (Li et al. 2008).

Microalgae shows a higher growth rate. This property makes possible to satisfy the huge demand on biofuels using limited land resources without causing shortage in biomass. Growing of microalgae requires lesser water in comparison to terrestial crops. The microalgae is able to tolerate high carbon dioxide content in gas streams. This would allow high-efficiency carbon dioxide alleviation. In comparison to conventional farming, microalgal farming could be more cost effective.

When microalgae are used for biofuel production, release of nitrous oxide could be reduced (Demirbas 2010d). The harvesting of algal biomass is costlier as the

Table 7.1 Biofuels from microalgae

Bioethanol
Vegetable oils
Biodiesel
Bio-oil
Bio-syngas
Bio-hydrogen

Table 7.2 Commonly used microalgae for biofuel production

Biofuel type	Microalgae
Bioethanol	Chlorococcum sp.
	Spirogyra
	Undaria pinnatifida
	Chlorella vulgaris
	Chlamydomonasreinhardtii
Biodiesel	Botryococcusbraunii
	Schizochytrium sp.
	Chlorella sp
Bio-oil	Corallinapilulifera
	Nannochloropsis sp.
	Dunaliella tertiolecta
	Chlorella sp.
Bio syngas	Chlorella vulgaris
	Chlorella vulgaris
	Cladophorafracta
	Botryococcusbraunii

Based on Raheem et al. (2015)

biomass concentration in the culture broth is lower because the size of algal cells is small and also there is limitation of light penetration (Soeder 1986). Microalgal farming facility involves higher capital costs and also it also requires thorough care in comparison to conventional agricultural farming. This is actually hindering the implementation of the biofuels from microalgae on a large scale (onlinelibrary. wiley.com; Wang and Yin 2018; Li et al. 2008).

Many microalgae can produce lipids under certain growth conditions (onlinelibrary.wiley.com). These lipids can be converted to biofuels used for transportation (Burlew 1953). The effects of environmental parameters on algal composition were studied (Spoehr and Milner 1949; Demirbas 2010d). The effects of varying concentration of nitrogen on the lipid and chlorophyll content of diatoms and *Chlorella* sp. have been reported (onlinelibrary.wiley.com). The fatty acid ranged between 10% and 30% dry cell weight for many green algae (Collyer and Fogg 1955). During silicon starvation, cellular lipids in the diatom increased (Werner 1966). In *Navicula pelliculosa*, the increase in lipid content was about 60% during a 14 h silicon starvation period (onlinelibrary.wiley.com). Composition of

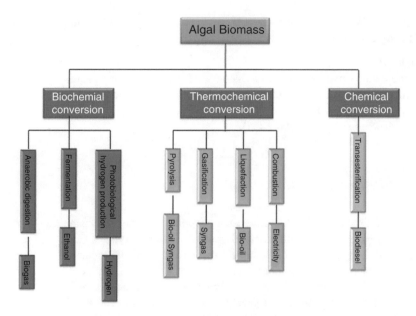

Fig. 7.1 Algal biomass conversion process for biofuel production. (Behera et al. 2015. Reproduced with permission)

fatty acids and lipids were also affected by many factors such as low temperatures and light (Demirbas 2010d; Coombs et al. 1967; Nichols 1965; Pohl and Wagner 1972; Rosenberg and Gouaux 1967; Ackman et al. 1968). "With the emergence of the oil embargo in 1970s, a search for alternative energy sources set the stage for an almost 20-year research effort devoted to biofuel production from algal lipids " (Demirbas 2010a).

Microalgal biomass can be converted into biofuel using several processes (www.mdpi.com) (Fig. 7.1) (Tsukahara and Sawayama 2005; Biomass R&D 2002; Show et al. 2017; John et al. 2011; Zhu et al. 2014, 2015; Behera et al. 2015). These processes are presented below (www.mdpi.com: d-nb.info):

Thermochemical Conversion
Thermochemical conversion process involves the thermal decomposition of organic matter in biomass to produce fuels. It includes the following:

- Direct combustion
- Gasification
- Thermochemical liquefaction
- Pyrolysis

This process overcomes the problems associated with biochemical conversion which are long retention time, higher cost of production and lower conversion efficiency by enzymes and microorganisms (econpapers.repec.org). Furthermore, this process can be integrated into the petroleum processing infrastructure (conservancy. umn.edu; econpapers.repec.org; www.assb.pl).

Biochemical Conversion
Biochemical conversion includes, alcoholic fermentation, anaerobic digestion and hydrogen production (pdfs.semanticscholar.org). These techniques can be used to produce bioethanol, biogas, bio-oil, biodiesel and hydrogen from microalgae (pdfs. semanticscholar.org; Pogaku 2015).

7.1 Biochemical Conversion

Microorganisms are able to convert the complex polymeric substances – proteins and carbohydrates – present in the biomass, to different type of fuels such as ethanol, butanol, hydrogen and methane (www.mdpi.com). Comparison of the basic properties of ethanol, butanol and gasoline are presented in Table 7.3. Biomethane is produced by anaerobic digestion of algal biomass (whole or spent), whereas alcoholic fermentation produces ethanol and butanol. These products from algal biomass can be used for combustion in power plants or in vehicles. For methane production different types of microalgal species have been examined (Table 7.4). For any type of biochemical conversion process, pre-treatment of the biomass is required (www.mdpi.com). Pre-treatment actually helps in breaking the rigid cell wall of microalgae and releases the cellular components efficiently. These are used in the later fermentation reactions (Show et al. 2017; Dudeja et al. 2012, www.mdpi. com). Diverse pretreatment methods and physic-chemical pretreatments have been used. Pretreatment methods used are physical, chemical and biological methods (Sharma and Arya 2017):

Physical methods – sonication, grinding, milling, pyrolysis.
Chemical methods – thermal treatment, treatment with acid or alkali
Biological methods – treatment with enzymes

Table 7.3 Comparison of the basic properties of ethanol, butanol and gasoline

Fuel	Ethanol	Butanol	Gasoline
Density at 15 °C (kg m^{-3})	795	810	750
Viscosity at 20 °C (mm^2 s^{-1})	1.52	3.64	0.4–0.8
Calorific value (MJ kg^{-1})	26.4	32.5	43.3
Octane number VM	108	96	95
Boiling point (°C)	78	118	30–190
Vapour pressure by Reida (kPa)	16.5	18.6	75
Oxygen content (% vol)	34.7	21.6	< 2.7

Based on Šebor et al. (2006), Mužíková et al. (2010), Hromádko et al. (2011) and Hönig et al. (2014)

Table 7.4 Yield of methane
from various feedstocks

Biomass	Methane yield ($m^3 \cdot kg^{-1}$)
Microalgae – ACAD model	0.54
Laminaria sp.	0.26–0.28
Gracilaria sp.	0.28–0.40
Sargassum sp.	0.12–0.19
Macrocystis	0.39–0.41
L. digitata	0.50
Ulva sp.	0.20
Water hyacinth	0.13–0.21
Sorghum	0.26–0.39
Poplar	0.23–0.32
Food waste	0.54

Based on Zhu et al. (2014)

The algal cell wall, is broken, the complex carbohydrates are hydrolyzed and the fermentable sugars are released (Behera et al. 2015; Varjani et al. 2018).

7.1.1 Anaerobic Digestion

The use of algae as a fuel for the production of methane gas was first proposed by Meier (1955). Oswald and Golueke (1960) further developed this idea. A conceptual technoeconomic engineering analysis of digesting microalgal biomass grown in large raceway ponds for methane production was introduced. The cost of conventional fuels started to rise speedily in the 1970s and the use of algae as a fuel source received renewed interest. Detailed design and engineering analysis of this concept was performed by Benemann et al. (1978). These systems could produce biogas competitively with projected fossil fuel prices (Demirbas 2010d; onlinelibrary. wiley.com). The organic matter gets decomposed in the absence of oxygen and electron acceptors such as nitrate, sulfate or ferric iron and produce methane and carbon dioxide. Microalgae can therefore be used for conversion into biomethane by anaerobic digestion process (Chen et al. 2016). Anaerobic process involves anaerobic bacteria, and the reactions between the substrate and the host are multifarious. This method is extensively used for the treatment of solid sewage, the organic fraction of municipal sewage and the digestion of manure. Anaerobic digestion can be used to convert the whole algal biomass into biogas in a single step (Leite et al. 2013; www. mdpi.com). Biogas is a valuable fuel. It is produced in digesters filled with the substrates such as dung or sewage. The digestion is continued for 10 days to a few weeks. Algal biomass can be used for production of biogas. There are several biogas installations in the world (Demirbas 2010d). These installations range from large-scale to small-scale fed with straw and green plant fuel that serve a few farms. Algae so far have not been used as a fuel. Certain types of macroalgal species such as

Macrocystis pylifera, and genera – *Cladophora, Chaetomorpha* and *Gracilaria, Sargassum, Laminaria, Ascophyllum, Ulva* have been studied as potential sources of methane (Filipkowska et al. 2008). Anaerobic digestion for biogas production does not appear to be viable in spite of the large seaweed biomass in different parts of the world (Gunaseelan 1997; Caliceti et al. 2002). Anaerobic digestion of the algal biomass generates methane, carbon dioxide along with ammonia. The remaining phosphorus and nitrogen compounds can be used as fertilizer in the algal process. Use of methane can further increase energy recovery from the process. The problems to be taken care of in the production of microalgae are that sodium in salt form can inhibit the anaerobic process with marine algae, although it has been suggested that suitable bacteria can adapt. The methane yield and digestion of algae can be enhanced by pre-treatment (physical or chemical) for rapturing the cell walls and making the organic matter present in the cells more accessible. In certain algae high content of nitrogen is present which produces higher amount of ammonia which inhibits the digestion process. By using a 'codigestion' process this problem can be solved. In this process, other waste having low nitrogen and high carbon content is added to the algal waste (www.fona.de; Takac̆ova et al. 2012).

In the anaerobic digestion process, following four stages are involved (www.mdpi.com):

– Hydrolysis
– Acidogenesis
– acetogenesis
– Methanogenesis.

In hydrolysis step, insoluble inorganic compounds are degraded and high-molecular weight compounds (polysaccharides, lipids, proteins and nucleic acids) get converted into soluble organic compounds (monosaccharides or amino acids.).

In the acidogenesis step, the resultant compounds get degraded and volatile fatty acids, ammonia and carbon dioxide are produced.

The products (organic acids and alohals) from the acidogenesis step are digested by acetogens into acetic acid, carbon dioxide and hydrogen. This process is called acetogenesis. It is affected by the hydrogen in the mixture.

Products from acetogenesis stage get converted into methane by using two types of methanogenic bacteria (Dote et al. 1994). One type converts acetate into methane and carbon dioxide, and the second group converts hydrogen and carbon dioxide into methane (Dote et al. 1994). Hydrolysis step is the main rate-limiting steps in anaerobic digestion (Lakaniemi et al. 2013). This step, is affected by the cell wall of microalgae. The structure of microalgal cell wall varies widely. Marine microalgae possess a thicker cell wall and fresh water microalgae have thinner wall. The presence of a rigid cell wall is quite challenging because it does not break easily. This affects the hydrolysis step, and the overall efficacy of the process. The cell wall of fresh water microalgae is less complex and does not need a severe treatment for anaerobic digestion process (www.mdpi.com).

When the ammonium salts are present, the anaerobic digestion process gets inhibited. Ammonium salt are present in two forms: the protonated form and the deprotonated form. Deprotonated ammonium has an inhibitory effect on the anaerobic digestion process due to its permeability through the cell wall. The distribution of both forms of ammonium salt is affected by the temperature and pH of the culture broth. At higher pH the production of ammonia is favored, whereas reducing the methanogenic activity reduces the generation of ammonia. This results in a drop in pH. Temperature affects the anaerobic digestion process. It changes the thermodynamics of biological processes and physical and physiochemical properties of the medium (www.mdpi.com). The mesophilic condition (30–38 °C) is suitable for mesophiles, whereas the thermophilic condition (49–57 °C) is suitable for thermophiles (Appels et al. 2008; Gonzalez-Fernandez et al. 2016). Thermophilic conditions are found to be suitable for anaerobic digestion. It provides the following benefits (www.mdpi.com)

- Better stabilization of waste
- More dewatering of sludge
- More production of methane
- Higher hydrolysis rate
- Reduced formation of foam
- Higher reduction of volatile organic compounds

Thermophilic conditions have certain disadvantages also (www.mdpi.com):

- High operating cost due to higher consumption of energy
- Sludge adaptation period is longer
- Stability is low
- Susceptibility to certain chemicals such as sodium, potassium or ammonium inhibition
- Generation of volatile fatty acids is high; the pH is affected

During the anaerobic digestion process, nutrients in the form phosphate and ammonia are produced. These nutrients can be reused as a substrate for growing microalgae (Hernandez et al. 2016; www.mdpi.com).

7.1.2 Alcoholic Fermentation

In this process, the carbohydrates present in the algal biomass are converted into alcohols (ethanol and butanol) by the action of different types of microorganisms. Most widely used microorganisms are bacteria, yeast and fungi. These microorganisms ferment the carbohydrates into ethanol and carbon dioxide under anaerobic conditions. The carbon dioxide produced in the fermentation process can be used into the ponds for growing microalgae. The fermentation can be presented by the following equation:

$$C_{12}H_{22}O_{11} \rightarrow C_6H_{12}O_6 + C_6H_{12}O_6$$
$$\text{Sucrose} \qquad\qquad \text{Glucose} \qquad \text{Fructose}$$

Microalgae has a complex cell wall. Due to this, microorganisms are not able to reach the carbohydrates. Therefore, pre-treatment is done. This step releases carbohydrates and converts them into monomers for conversion by microorganisms. Various pre-treatment methods have been used. These methods are – physical or mechanical, thermal, chemical and enzymatic. A combined thermal and chemical pre-treatment process is found to be the most commonly used method for releasing sugars. This involves hydrolyzing algal biomass at high temperature in the presence of mild acid or alkali. This pre-treatment method gives a high recovery of simple sugars with a higher efficiency because microalgae lack lignin (Mendez et al. 2014). Under certain nutrient starvation conditions (nitrogen and phosphorus), *Chlorella, Chlamydomonas, Scenedesmus, Dunaliella, Tetraselmis* and *Spirulina* possess a higher carbohydrate content (>40% by weight) (Wang et al. 2015). In green algae and cyanobacteria, the carbohydrates are associated with their cells as storage or structural polysaccharides (www.mdpi.com). Carbohydrates produces glucose and xylose as the major sugars. Other sugars arabinose, mannose and galactose are also produced. Fermenting microorganisms are able to utilize these sugars efficiently (Wang et al. 2015). The most commonly used yeast for production of ethanol on an industrial scale is *Saccharomyces cerevisiae* (www.mdpi.com). This yeast is able to utilize only hexose sugars and produces ethanol, but pentose sugars are not utilized. Higher salt concentration in the hydrolysate of marine microalgae affects the fermentation of *S. cerevisiae*. The biomass of *Arthrospira platensis* was used for ethanol production with a salt stress–adapted *S. cerevisiae* by Markou et al. (2013). Acid treatment and thermal treatments were combined and used; an increase in the acid strength and temperature increased the reducing sugars. Optimal results were obtained by the use of low-strength acids at a higher temperature. High ethanol yield (16.5%) was achieved for *A. platensis* biomass treated with both 0.5 N sulphuric acid- and nitric acid (John et al. 2011). Acidic and enzymatic methods were used for pretreatment of *Chlorella vulgaris* FSP-E. Pretreated biomass was then used for production of ethanol using *Zymomonas mobilis* (Markou et al. 2013). This bacterium produces higher ethanol yield. It does not require controlled addition of oxygen during fermentation. It also shows high utilization of substrate and is able to tolerate higher ethanol concentration up to 120 g/L. The possibility for genetic manipulation is also high (www.mdpi.com). Hydrolysis of *C. vulgaris* FSP-E with enzyme produced about 90.4% glucose and with mild acid hydrolysis about 93.6% glucose was obtained. Use of super high frequency (SHF) with enzymatic hydrolysate produced about 79.9% of the theoretical yield, whereas SHF with acid hydrolysate produced 87.6% of the theoretical yield (Markou et al. 2013; www.mdpi.com).

The process for bioethanol production from microalgae is presented in Fig. 7.2 (archive.org).

Biobutanol, can also be produced by fermentation using microalgal hydrolysate as a substrate (Show et al. 2017; www.mdpi.com). It is an alternative fuel that is completely interchangeable and compatible with a particular conventional (typically

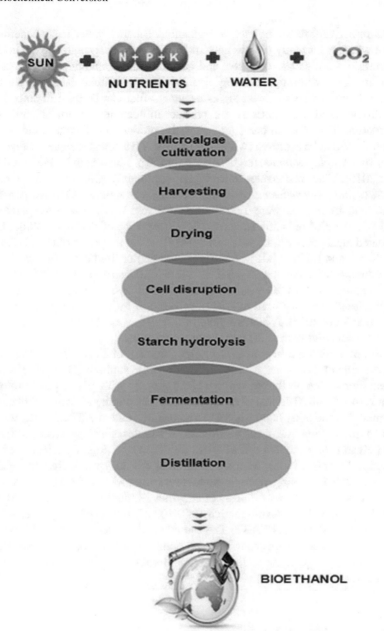

Fig. 7.2 Process for bioethanol production from microalgae. (Behera et al. 2015. Reproduced with permission)

petroleum-derived) fuel to be used with gasoline. Butanol shows many benefits over ethanol which is currently used as an additive in gasoline. It can be mixed in higher concentrations and is less corrosive and generates more energy per unit mass and can be transported through existing pipelines (www.yesitekhob.com; www.news-wise.com/.../researchers-convert-algae-to-butanol-fuel-can-be-used-in-auto...)

Hydrogen is also obtained as a byproduct in acetone–butanol–ethanol (ABE) fermentation. Therefore in one single step, the production of both liquid and gaseous biofuels can be achieved (www.mdpi.com). A wastewater algae – *Clostridium saccharoperbutylacetonicum* has been used for the production of biobutanol (Castro et al. 2015). The wastewater microalgae are mainly *Chlorella, Micromonas, Ankistrosdemus, Scenedesmus and Chlamydomonas species*. Optimal conditions for the acid hydrolysis were 1.0 M sulphuric acid for 2 h at a temperature of 80–90 °C. Yield of reducing sugar was 166.1 g per kg of dry algae. With 10% of pre-treated algae, 5.23 g/L of ABE and 3.74 g/L of butanol were obtained (Ho et al. 2013; Singh and Rathore 2017). SMAB obtained after lipid extraction can be used for production of butanol and ABE fermentation(Castro et al. 2015). SMAB of *Chlorella sorokiniana* CY1 was subjected to acidic hydrolysis under milder reaction conditions and was used in ABE fermentation by *C. acetobutylicum* (Castro et al. 2015; Yang 2015). A butanol yield of 3.86 g/L was obtained from microalgal sugars at a concentration of 300 g/L (www.mdpi.com).

ABE fermentation using wastewater algal biomass of *Clostridium saccharoperbutylacetonicum* N1–4 as a carbon source was studied (Ellis et al. 2012). Fermentations were performed using a batch process with 10% algae as substrate. About 2.74 g/L of ABE was obtained when algae pretreated with acid/base was fermented. On the other hand when 1% glucose was added 7.27 g/L ABE was produced (digitalcommons.usu.edu). Addition of cellulases and xylanases to the pretreated algae produced 9.74 g/L of ABE. Addition of 1% glucose increased ABE production by about 160%, whereas addition of the enzymes increased ABE production by 250% in comparison to production from pretreated algae with no addition of enzymes and sugar. Addition of enzymes resulted in the highest ABE yield of 0.311 g/g and volumetric productivity of 0.102 g/L h. Use of non-pretreated algae produced about 0.73 g/L of ABE. The use of novel strategies for producing these value added products from renewable and abundantly available feedstock such as algae could have a great impact in stimulating domestic energy economies (digitalcommons.usu.edu).

7.1.3   Hydrogen Production

Hydrogen is a promising future transportation fuel. The conventional methods of producing hydrogen are energy intensive, costly and not environmentally friendly. Algal biomass is being considered as an attractive raw material for biofuel production. Open-air systems and photobioreactors are being studied for hydrogen production from algal biomass (Behera et al. 2015). Open-air systems are inexpensive and simpler to operate. Careful control of culture conditions are possible in Photobioreactors. Immobilization of algal cell on various matrixes results in substantial increase in the reactor productivity (Sharma and Arya 2017; www.ncbi.nlm.nih. gov).

Several algal species under certain conditions show potential to produce hydrogen (Li and Fang 2007; Sharma and Arya 2017). However, there is a need to overcome the following technical barriers before using microalgae as a viable feedstock (Radakovits et al. 2010; Sharma and Arya 2017):

- Developing methods involving low energy for harvesting the algal cells
- Problems in producing biomass on a large scale
- Presence of offensive species in large-scale ponds
- Low penetration of light in dense cultures
- Lack of cost-effective techniques for extraction

"Mixed bacterial culture (photosynthetic anaerobic bacteria) provides a method for use of a variety of resources for biohydrogen- production" (Miyake 1990).

Hydrogen can be produced from algal biomass by using anaerobic bacteria. The fermentation is carried out in dark (Show et al. 2017). Also, hydrogen gas can be produced directly from microalgae by the water-splitting activity of the photosystems involved in photosynthesis (www.mdpi.com). "In these light-dependent systems, the water-splitting activity of photosystem II is supplied with electrons by the light-excited photosystem I or by the intracellular plastoquinone pool derived from the metabolism of intracellular carbohydrates. Either way, photosystem II is the water-splitting component and the resultant electrons are transferred to protons to produce hydrogen, catalyzed by the enzyme hydrogenase" (Eroglu and Melis 2016; www.mdpi.com). In green algae and cyanobacteria, hydrogenase enzyme is broadly distributed. Hydrogenases present in green algae are able to produce hydrogen only, whereas in cyanobacteria, the hydrogenases are bidirectional and use hydrogen released during fixation of nitrogen. Under certain physiological conditions, production of hydrogen by green algae and cyanobacteria occurs temporarily (Eroglu and Melis 2016). In green algae, particularly in *Chlamydomonas reinhardtii*, a continuous hydrogen production has been obtained using a two-phase method involving deprivation of macronutrient. "Sulfur deprivation of, the microalgal cultures after exponential growth resulted in a partial inactivation of photosystem II, inducing cellular respiration, reducing its activity, establishing anoxia and obtaining continued production of hydrogen for a period of five to seven days" (Melis et al. 2000). This method was found to work well. The resultant algal biomass needs to be taken care of. There are several benefits of using microalgae for production of hydrogen but commercialization of this approach appears to be a long-term objective. More inputs are required in terms of operation feasibility. (www.mdpi.com).

Factors affecting the cost of hydrogen production by microalgae are the cost of hydrogen storage facilities and the photobioreactor which guarantee continuous supply of hydrogen during the night or during the cloudy periods (Demirbas 2010d).

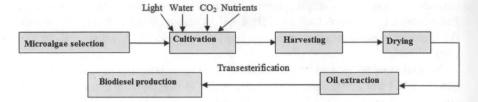

Fig. 7.3 Different stages of production of microalgal diesel. (Based on Saifullah et al. 2014)

7.1.4   Biodiesel Production

The problems in production of biodiesel is in selecting an algal strain which contain high amount of lipid and grows fast, does not pose any difficulty in harvesting. The cultivation system should be cost effective (Demirbas and Demirbas 2011). Use of genetic engineering, screening and selection methods, new design and materials for cultivating algae in closed bioreactor systems show promise (Schenk et al. 2008; microbewiki.kenyon.edu).

Microalgae have been also explored for the production of biodiesel. High yield of biodiesel are obtained (Pragya et al. 2013; www.mdpi.com). Figure 7.3 shows different stages of production of microalgal diesel. Vegetable oil, animal fat and used edible oil are used as substrates for commercial production of biodiesel (Barnwal and Sharma 2005). For production of bioethanol, sugar cane, sugar beets and corn starch, are used and for the production of biodiesel, oilseed rape and palm are used (Clark and Deswarte 2015). Industry is exploring microalgae as a new biodiesel source as the demand for biodiesel is increasing. The profile of fatty acid from algal oil is compatible with the production of biodiesel (Gouveia and Oliveira 2009). After drying and processing, the lipids or oils are extracted from microalgae for transesterification. For alcoholysis, ethanol, methanol, propanol, butanol and amyl alcohol are used. Transesterification is performed with and without catalyst. Both chemical and enzymatic catalysts have been used. The product obtained from the transesterification of fatty acid methyl esters can be used as a fuel in diesel engines (Fukuda et al. 2001). Biodiesel yield from different sources has been reported (Zhu 2015). About 58,700 L of oil can be produced per hectare of microalgae which is two times higher compared to the oil produced from other crops (Chisti 2007; www.mdpi.com).

7.2   Thermochemical Conversion

Main thermochemical conversion processes include – Gasification, Liquefaction and Pyrolysis (Demirbas 2010b, c). Algal cells are freeze dried and sonicated, and hydrocarbons are separated by extraction with organic solvent. These methods are expensive to use and therefore not suitable for large-scale separation. Liquefaction method is found to be effective for separating hydrocarbons from algal cells containing high moisture. This process can convert wet biomass to liquid fuel at 575 K and 10 MPa using sodium carbonate as catalyst (Demirbas 2007). The separation of liquid oil is easier (Ogi et al. 1990). Several researchers have reported the practicality of producing liquid fuel via thermochemical liquefaction of different types of microalgae (Sawayama et al. 1994, 1999; Peng et al. 2000, 2001; Tsukahara and Sawayama 2005; Demirbas 2010c; Singh and Kalia 2017; Yan et al. 2009).

The moisture content in algae is high so drying is needed. This process requires high energy for heating (Yang et al. 2004; Prabandono and Amin 2015). A new process – low temperature catalytic gasification of the microalgae has been recommended. This process produces combustible gas. Bio-oil is produced in the pyrolysis process (Demirbas 2010d; Minowa and Sawayama 1999; Chisti 2006; Cohen et al. 1991).

7.2.1   Gasification

This process produces syngas from different types of feedstocks which includes microalgae also (Du 2013). In this process, algal biomass is reacted with air at temperatures in the range of 800–1000 °C in the presence of low amount of oxygen. The syngas is produced which is a mixture of carbon monooxide, hydrogen, carbon dioxide, nitrogen dioxide and methane. (Demirbas 2001; www.mdpi.com). A lot of research has been conducted on terrestrial biomass but there are not too many publications and reports on gasification of microalgae. Most of these studies pertain to hydrothermal gasification (conservancy.umn.edu). Hydrothermal technologies deal with physical and chemical transformations which are performed at high-pressure water and high temperature (Peterson et al. 2008). Pressure is used in the range of 5–40 MPa and temperature in the range of 200–600 °C (conservancy.umn.edu).

Many benefits are seen in case of wet biomass. Water is heated in the pressurized form therefore the energy input for removing water by evaporation is removed. Gasification processes is performed with and without catalyst at high temperatures of more than 370 °C. Hydrothermal gasification of *Chlorella vulgaris* was studied by Chakinala et al. (2010) at 400–700 °C. Gasification was found to increase with higher temperature, longer residence times and lower algae concentrations. Complete gasification occurred at 700 °C in the presence of Ru/TiO2 catalysts. In case of *Spirulina platensis*, 60–70% of the heating value was recovered as methane at 399–409 °C in presence of Ru/C or Ru/ZrO2 catalyst (Stucki et al. 2009; conservancy.umn.edu).

Syngas is produced in the gasification process. Its calorific-value is low and can be directly used as fuel in turbines and engines. It can be also used for methanol synthesis in chemical conversion processes (www.mdpi.com; Hirano et al. 1998). Different microalgae have been studied for gasification. Production of methanol from *Spirulina* sp. was studied by Hirano et al. (1998). The algal slurry was continuously pumped to the reactor and was oxidized at 800–1000 °C. The syngas was produced which contained hydrogen, carbon dioxide, carbon monooxide, and methane. Ethylene, nitrogen, and oxygen were present in trace amount (www.mdpi. com). When the temperature was increased, the concentration of hydrogen increased and the concentrations of carbon monooxide, carbon dioxide, and methane decreased. Optimum temperature producing the highest yield of methanol was found to be 1000 °C (Hirano et al. 1998). The slurry of microalgae with a moisture content of around 15% can be used for gasification. Algae having high moisture around 40% has been also used (Raheem et al. 2015). In gasification, a high temperature of 1300 °C is required. With catalyst, the reaction temperature can be reduced (Raheem et al. 2015; www.mdpi.com).

Ebadi et al. (2018) examined the gasification of *Cladophora glomerata L.* with steam with catalysts (alkali and alkaline-earth metal) for increasing the syngas yield and reducing the tar content by cracking and reforming of condensable fractions. The commercial catalysts used were sodium hydroxide, sodium phosphate, potassium bicarbonate, and magnesium oxide. The gasification studies were conducted on a lab scale, biomass gasification unit. Sodium hydroxide showed a strong potential for hydrogen production. It also showed certain other advantages of char conversion and tar destruction which improved, calorific value of produced syngas. Increasing the temperature from 700 to 900 °C, reduced the tar content but the hydrogen yield increased. Increasing the ratio of steam/biomass substantially increased tar destruction and the hydrogen yield. The particle size (0.5–2.5 mm) did not play a significant role in the process (Ebadi et al. 2018).

7.2.2 Hydrothermal Liquefaction (HTL)

This is a thermal depolymerization process. At high pressure and moderate temperature, wet biomass is converted into crude oil or bio-oil (Leng et al. 2018). The energy density of this oil is high and the heating value is lower (33.8–36.9 MJ/kg). Catalysts (homogeneous and/or heterogeneous) are used for improving the yield and the product quality. Carbon and hydrogen of biomass, peat are converted by thermo-chemical process into hydrophobic compounds having low viscosity and high solubility. The fuel are used in heavy engines and can be upgraded to transportation fuels also (en.wikipedia.org).

Liquefaction is conducted directly or indirectly. In direct liquefaction rapid pyrolysis is performed and liquid tars, oils and/or condensable organic vapors are produced (Demirbas 2010d). In indirect liquefaction catalysts are used for converting products of pyrolysis or gasification into liquid products. Sodium carbonate and potassium carbonate, are able to catalyse the hydrolysis of cellulose and hemicellulose into smaller fragments. "The degradation of biomass into smaller products mainly proceeds by depolymerization and deoxygenation " (Demirbas 2000).

More papers have been published on HTL of algae in comparison with gasification (conservancy.umn.edu). This process directly produces liquid fuels. It is performed at 250–350 °C, which is referred to as the subcritical water condition. The quality of bio-oil and yield is a function of temperature, biomass loading, residence time and presence of catalysts. The bio-oil yield ranged from 10% to 50%. The heating value was slightly lower compared to petroleum crude oil (Biller and Ross 2011; Duan and Savage 2011; Zou et al. 2009). In HTL, a large amount of water is present. Therefore, bio-oil is emulsified and large amount of organic solvent is required for extraction.

With *Dunaliella tertiolecta* with a moisture content of 78.4 wt.%, an oil yield of about 37% was achieved by using direct hydrothermal liquefaction at around 575 K and 10 MPa (Minowa et al. 1995; Demirbas 2010d). The viscosity of oil was 150–330 Mpas and heating value was 36 MJ/kg. With *B. braunii* (a colony-forming microalga), a higher amount of oil was obtained. The yield was 57–64 wt.% at 575 K. The quality of oil was comparable to petroleum oil. Maximum recovery of hydrocarbons was >95% at 575 K (Banerjee et al. 2002; Demirbas 2010b).

7.2.3   Hydrothermal Carbonization

HTC process is also known as wet torrefaction (conservancy.umn.edu; Funke and Ziegler 2010). This is a chemical process converts organic compounds into structured carbons. It is combined dehydration and decarboxylation of a fuel for increasing its carbon content in order to obtain a higher calorific value. It can be also referred to as aqueous carbonization at high pressure and temperature and can be used for producing the following with release of energy (en.wikipedia.org):

– Synthesis gas
– Liquid petroleum precursors
– Nanostructured carbons
– Brown coal substitute
– Humus from biomass

Milder conditions are used in this process in comparison to liquefaction and hydrothermal gasification. Biomass is heated at a temperature of about 200 °C in water under pressure for a certain time (conservancy.umn.edu). HTC is more suitable for the conversion of wet biomass. In algae, water needed is already present in the biomass. HTC of different algae was examined under different conditions (Heilmann et al. 2010). Different amount of carbonization was observed. The char products of bituminous quality were obtained. Comparison of energy input and output for combustion of green alga *Chlamydomonas reinhardtii* and its char product was made. Energy of 12.01 MJ/kg was obtained when the algal char was combusted. The energy loss was 5.27 MJ/kg for starting algae as HTC process significantly reduces the energy needed for drying of algae. Fatty acids were absorbed on the char in high yield. The fatty acids in the char could be extracted and converted into liquid fuels after extraction (Heilmann et al. 2011; conservancy.umn.edu).

7.2.4   Pyrolysis

Pyrolysis has been studied by several researchers (Goyal et al. 2008). Depending upon the process, the pyrolysis of algal biomass can produce all types of biofuels (Miao and Wu 2004; Demirbas 2006; Miao et al. 2004). This process is performed at high temperature (300–700 °C) in the absence of air. Biomass gets converted into bio-oil, biochar and syngas (Bridgwater and Peacocke 2000; www.mdpi.com; conservancy.umn.edu; bala et al. 2014; Fermoso et al. 2017). The properties and yield depend on the following factors:

– Temperature
– Retention time
– Heating rate
– Catalyst

For pyrolysis, microalgae could be a potential substrate as the bio-oils produced from microalgae are found more stable in comparison to bio-oils obtained from lignocellulosic biomass (Raheem et al. 2015). Temperature has an effect on this process. At optimum temperature, an oil yield of 55.3% has been obtained with *Chlorella prothothecoides* (Demirbas 2006; www.mdpi.com).

According to the residence time of vapors and the heating rate, pyrolysis can be classified into slow pyrolysis and fast pyrolysis. Fast pyrolysis produces more than 50% of liquid fuel and is conducted at a higher temperature (500 °C). Slow pyrolysis is performed at a lower temperature (400 °C). In this process around 35% of biochar along with, 35% of gas, and 30% of water are produced (Bridgwater 2007). The yield of oil from this process is higher by 3.4 times compared to phototrophic cultivation (Miao and Wu 2004). For production of charcoal, slow pyrolysis having a residence time from minutes to hours has been conducted for several years. Research is focusing on the optimization of the process for liquid bio-oil. This can be easily, stored and transported. Its upgradation to high quality fuels can also be performed. In industrial boiler, light and heavy fuel oils can be replaced by bio-oil for production of heat (Mohan et al. 2006). But, due to the presence of high level of oxygen, bio-oil becomes unstable and cannot be directly used as transportation fuels (www.mdpi.com; conservancy.umn.edu).

Most of conventional pyrolysis processes use fluidized bed and fixed bed reactors. In these reactors heating is done by heated surface and sands, etc. (conservancy.umn.edu). Microwave heating has been also explored (Meier and Faix 1999; Mohan et al. 2006; Czernik and Bridgwater 2004; Wan et al. 2009; Yu et al. 2008). Electromagnetic waves pass through material, the molecules oscillate and heat is produced. There are many benefits of MAP over the conventional pyrolysis. These are presented below (conservancy.umn.edu):

– Can be used for large particles of biomass as the heating is uniform
– The heating value of syngas is higher as it is not diluted by carrier gas used in some types of pyrolysis. Combustion of syngas can be done to provide in-situ electricity for generation of microwaves.

- Products are clean
- Technology can be scaled up easily

Several papers have been published on the pyrolysis of lignocellulosic feedstock (Yu et al. 2017). There is not much information on production of bio-oil from algae using the pyrolysis process. Fast pyrolysis of *Microcystis areuginosa* and *Chllorella protothecoides* was peformed at temperature of 500 °C (Miao et al. 2004). Bio-oil yield of 18% was obtained with *Chllorella protothecoides* and 24% with *Microcystis areuginosa*. Bio-oil contained a lower oxygen content, higher carbon and nitrogen content as compared to wood bio-oil. Growing *Chllorella protothecoides* in a heterotrophic manner, increased bio-oil yield to 57.9% having a heating value of 41 MJ/kg (Miao et al. 2004). Microalgae as the third generation of biofuel has become a hot research topic in the recent years, so pyrolysis as a potential conversion method is attracting attention for production of biofuel.

Slow pyrolysis of *Nannochloropsis* sp. without and with catalyst (HZSM-5) has been studied (Pan et al. 2010; Fermoso et al. 2017). Bio-oil rich in aromatic hydrocarbons was obtained. Pyrolytic products obtained from algae separate into two phases (conservancy.umn.edu). Bio-oil remains in the top phase (Campanella et al. 2012; Jena and Das 2011; Du et al. 2012). Bio-oil from algae have higher heating values (31–36 MJ/kg). These values are higher in comparison to those achieved with lignocellulosic biomass. In comparison to bio-oil produced by hydrothermal liquefaction, bio-oil from algae contains more low boiling compounds. This oil consists of lower molecular weights compounds. It matches with Illinois shale oil and is suitable for petroleum fuel replacement (Vardon et al. 2012; Jena and Das 2011). The nitrogen content in the bio-oil is high due to the presence of proteins in the algae. This results in NOx emissions during combustion. When co-processed in crude oil refineries, there is deactivation of acidic catalysts also. Bio-oil from algae has a lower oxygen content, higher heating value and a pH value higher than 7. Removal of oxygen and nitrogen in the bio-oil is required before using it in drop-in fuels (conservancy.umn.edu).

References

Ackman RG, Tocher CS, McLachlan J (1968) Marine phytoplankter fatty acids. J Fish Res Board Can 25.1603–1620

Appels L, Baeyens J, Degrève J, Dewil R (2008) Principles and potential of the anaerobic digestion of waste-activated sludge. Prog Energy Combust Sci 34:755–781

Bala K, Kumar R, Deshmukh D (2014) Perspectives of microalgal biofuels as a renewable source of energy. Energy Convers Manag 88:1228–1244

Banerjee A, Harma RS, Chisti Y, Banerjee UC (2002) *Botryococcus braunii*: a renewable source of hydrocarbons and other chemicals. Crit Rev Biotechnol 22:245–279

Barnwal B, Sharma M (2005) Prospects of biodiesel production from vegetable oils in India. Renew Sust Energy Rev 9:363–378

Behera S, Singh R, Arora R, Sharma NK, Shukla M, Kumar S (2015) Scope of algae as third generation biofuels, Frontiers in bioengineering and biotechnology. Mar Biotechnol 90(2):1–13

Benemann JR, Pursoff P, Oswald WJ (1978) Engineering design and cost analysis of a large-scale microalgae biomass system. NTIS#H CP/T1605–01 UC-61. US Department of Energy, Washington DC

Biller P, Ross AB (2011) Potential yields and properties of oil from the hydrothermal liquefaction of microalgae with different biochemical content. Bioresour Technol 102:215–225

Biomass R&D (2002) Technical advisory committee. Roadmap for biomass technologies in the United States, Washington, DC, USA. Available online: www.bioproducts-bioenergy.gov/pdfs/FinalBiomassRoadmap.pdf

Bridgwater A (2007) IEA bioenergy 27th update. Biomass pyrolysis, biomass and bioenergy, vol 31. Pergamon-Elsevier Science Ltd., England

Bridgwater AV, Peacocke GVC (2000) Fast pyrolysis processes for biomass. Renew Sust Energ Rev 4:1–73

Burlew S (1953) Algal culture: from laboratory to pilot plant (publication no. 600). Carnegie Institution of Washington, Washington, DC

Caliceti M, Argese E, Sfriso A, Pavoni B (2002) Heavy metal contamination in the seaweeds of the Venice lagoon. Chemosphere 47:443–454

Campanella A, Muncrief R, Harold MP, Griffith DC, Whitton NM, Weber RS (2012) Thermolysis of microalgae and duckweed in a CO_2-swept fixed-bed reactor: bio-oil yield and compositional effects. Bioresour Technol 109:154–162

Castro YA, Ellis JT, Miller CD, Sims RC (2015) Optimization of wastewater microalgae saccharification using dilute acid hydrolysis for acetone, butanol, and ethanol fermentation. Appl Energy 2015(140):14–19

Chakinala AG, Brilman DWF, van Swaaij WPM, Kersten SRA (2010) Catalytic and non-catalytic supercritical water gasification of microalgae and glycerol. Ind Eng Chem Res 49:1113–1122

Chen CY, Chang HY, Chang JS (2016) Producing carbohydrate-rich microalgal biomass grown under mixotrophic conditions as feedstock for biohydrogen production. Int J Hydrog Energy 41:4413–4420

Chisti Y (2006) Microalgae as sustainable cell factories. Environ Eng Manag J 5:261–274

Chisti Y (2007) Biodiesel from microalgae. Biotechnol Adv 25:294–306

Clark JH, Deswarte F (2015) Introduction to chemicals from biomass. Wiley, Hoboken

Cohen E, Koren A, Arad SM (1991) A closed system for outdoor cultivation of microalgae. Biomass Bioenergy 1:83–88

Collyer DM, Fogg GE (1955) Studies of fat accumulation by algae. J Exp Bot 6:256–275

Coombs J, Darley WM, Holm-Hansen O, Volcani BE (1967) Studies on the biochemistry and fine structure of silica shell formation in diatoms. Chemical composition of Navicula pelliculosa during silicon starvation. Plant Physiol 42:1601–1606

Czernik S, Bridgwater AV (2004) Overview of applications of biomass fast pyrolysis oil. Energy Fuel 18:590–598

Demirbas A (2000) Mechanisms of liquefaction and pyrolysis reactions of biomass. Energy Convers Manag 41:633–646

Demirbas A (2001) Biomass resource facilities and biomass conversion processing for fuels and chemicals. Energy Convers Manag 42:1357–1378

Demirbas A (2006) Oily products from mosses and algae via pyrolysis. Energy Source Part A 28:933–940

Demirbas A (2007) Progress and recent trends in biofuels. Prog Energy Combust Sci 33:1–18

Demirbas A (2010a) Thermochemical processes. In: Biorefineries. Green energy and technology. Springer, London

Demirbas A (2010b) Energy from algae, green energy and technology. Springer, London

Demirbas MF (2010c) Microalgae as a feedstock for biodiesel. Energy Educ Sci Technol Part A 25:31–43

Demirbas A (2010d) Use of algae as biofuel sources. Energy Convers Manag 51(12):2738–2749

Demirbas A, Demirbas F (2011) Importance of algae oil as a source of biodiesel. Energy Convers Manag 53:163–170. https://doi.org/10.1016/j.enconman.2010.06.055

Dote Y, Sawayama S, Inoue S, Minowa T, Yokoyama SY (1994) Recovery of liquid fuel from hydrocarbon-rich microalgae by thermochemical liquefaction. Fuel 73:1855–1857

Du Z (2013) Thermochemical conversion of microalgae for biofuel production. Published doctoral dissertation, University of Minnesota, Twin Cities

Du Z, Mohr M, Ma X, Cheng Y, Lin X, Liu Y, Zhou W, Chen P, Ruan R, Bioresource Technology (2012) Hydrothermal pretreatment of microalgae for production of pyrolytic bio-oil with a low nitrogen content. Bioresour Technol 120:13–18

Duan PG, Savage PE (2011) Hydrothermal liquefaction of a microalga with heterogeneous catalysts. Ind Eng Chem Res 50:52–61

Dudeja S, Bhattacherjee AB, Chela-Flores J (2012) Antarctica as model for the possible emergence of life on Europa. In: Hanslmeier A, Kempe S, Seckbach J (eds) Life on earth and other planetary bodies. Cellular origin and life in extreme habitats and astrobiology. Springer, Dordrecht

Ebadi AG, Hisoriev H, Zarnegar M, Ahmadi H (2018) Hydrogen and syngas production by catalytic gasification of algal biomass (*Cladophora glomerata* L.) using alkali and alkaline-earth metals compounds. Environ Technol 2:1–7. https://doi.org/10.1080/09593330.2017.1417495

Ellis JT, Hengge NN, Sims RC, Miller CD (2012) Acetone, butanol, and ethanol production from wastewater algae. Bioresour Technol 111:491–495

Eroglu E, Melis A (2016) Microalgal hydrogen production research. Int J Hydrog Energy 41:12772–12798

Fermoso J, Coronado JM, Serrano DP, Pizarro P (2017) Pyrolysis of microalgae for fuel production. In: Gonzalez-Fernandez C, Muñoz R (eds) Microalgae-based biofuels bioprod. Woodhead Publishing/Elsevier, Duxford, pp 259–282

Filipkowska A, Lubecki L, Szymczak-Żyła M, Kowalewska G, Żbikowski R, Szefer P (2008) Utilisation of macroalgae from the Sopot beach (Baltic Sea). Oceanologia 50:255–273

Fukuda H, Kondo A, Noda H (2001) Biodiesel fuel production by transesterification of oils. J Biosci Bioeng 2001(92):405–416

Funke A, Ziegler F (2010) Hydrothermal carbonization of biomass: a summary and discussion of chemical mechanisms for process engineering. Biofuels Bioprod Biorefin 4:160–177

Gonzalez-Fernandez C, Mandy A, Ballesteros I, Ballesteros M (2016) Impact of temperature and photoperiod on anaerobic biodegradability of microalgae grown in urban wastewater. Int Biodeterior Biodegrad 106:16–23

Gouveia L, Oliveira AC (2009) Microalgae as a raw material for biofuels production. J Ind Microbiol Biotechnol 36:269–274

Goyal H, Seal D, Saxena R (2008) Bio-fuels from thermochemical conversion of renewable resources: a review. Renew Sust Energ Rev 12:504–517

Gunaseelan VN (1997) Anaerobic digestion of biomass for methane production: a review. Biomass Bioenergy 13:83–114

Heilmann SM, Davis HT, Jader LR, Lefebvre PA, Sadowsky MJ, Schendel FJ, von Keitz MG, Valentas KJ (2010) Hydrothermal carbonization of microalgae. Biomass Bioenergy 34:875–882

Heilmann SM, Jader LR, Harned LA, Sadowsky MJ, Schendel FJ, Lefebvre PA, von Keitz MG, Valentas KJ (2011) Hydrothermal carbonization of microalgae II. Fatty acid, char, and algal nutrient products. Appl Energy 88:3286–3290

Hernandez D, Riano B, Coca M, Solana M, Bertucco A, Garcia-Gonzalez MC (2016) Microalgae cultivation in high rate algal ponds using slaughterhouse wastewater for biofuel applications. Chem Eng J 285:449–458

Hirano A, Hon-Nami K, Kunito S, Hada M, Ogushi Y (1998) Temperature effect on continuous gasification of microalgal biomass: theoretical yield of methanol production and its energy balance. Catal Today 45:399–404

Ho SH, Huang SW, Chen CY, Hasunuma T, Kondo A, Chang JS (2013) Bioethanol production using carbohydrate-rich microalgae biomass as feedstock. Bioresour Technol 2013(135):191–198

Hönig V, Kotek M, Mařík J (2014) Use of butanol as a fuel for internal combustion engines. Agron Res 12(2):333–340

Hromádko J, Hromádko J, Miler P, Hönig V, Štěrba P (2011) The use of bioethanol in internal combustion engines. Chemické listy 105(2):122–128 (in Czech)

Jena U, Das KC (2011) Comparative evaluation of thermochemical liquefaction and pyrolysis for bio-oil production from microalgae. Energy Fuel 25:5472–5482

John RP, Anisha GS, Nampoothiri KM, Pandey A (2011) Micro and microalgal biomass: a renewable source for bioethanol. Bioresour Technol 102:186–193

Lakaniemi AM, Tuovinen OH, Puhakka JA (2013) Anaerobic conversion of microalgal biomass to sustainable energy carriers – a review. Bioresour Technol 135:222–231

Leite GB, Abdelaziz AE, Hallenbeck PC (2013) Algal biofuels: challenges and opportunities. Bioresour Technol 145:134–141

Leng L, Li J, Wen Z, Zhou W (2018) Use of microalgae to recycle nutrients in aqueous phase derived from hydrothermal liquefaction process. Bioresour Technol 256:529–542

Li CL, Fang HHP (2007) Fermentative hydrogen production from wastewater and solid wastes by mixed cultures. Crit Rev Environ Sci Technol 37:1–39

Li Y, Horsman M, Wu N, Lan CQ, Dubois-Calero N (2008) Biofuels from microalgae. Biotechnol Prog 24(4):815–820

Markou G, Angelidaki I, Nerantzis E, Georgakakis D (2013) Bioethanol production by carbohydrate-enriched biomass of Arthrospira (Spirulina) platensis. Energies 2013(6):3937–3950

Meier RL (1955) Biological cycles in the transformation of solar energy into useful fuels. In: Daniels F, Duffie JA (eds) Solar energy research. University of Wisconsin Press, Madison, pp 179–183

Meier D, Faix O (1999) State of the art of applied fast pyrolysis of lignocellulosic materials – a review. Bioresour Technol 68:71–77

Melis A, Zhang L, Forestier M, Ghirardi ML, Seibert M (2000) Sustained photobiological hydrogen gas production upon reversible inactivation of oxygen evolution in the green alga *Chlamydomonas reinhardtii*. Plant Physiol 122:127–136

Mendez L, Mahdy A, Ballesteros M, Gonzalez-Fernandez C (2014) Methane production of thermally pretreated *Chlorella vulgaris* and *Scenedesmus* sp. biomass at increasing biomass loads. Appl Energy 129:238–242

Miao XL, Wu QY (2004) High yield bio-oil production from fast pyrolysis by metabolic controlling of *Chlorella protothecoides*. J Biotechnol 110:85–93

Miao XL, Wu QY, Yang CY (2004) Fast pyrolysis of microalgae to produce renewable fuels. J Anal Appl Pyrolysis 71:855–863

Minowa T, Sawayama S (1999) A novel microalgal system for energy production with nitrogen cycling. Fuel 78:1213–1215

Minowa T, Yokoyama SY, Kishimoto M, Okakura T (1995) Oil production from algal cells of Dunaliella tertiolecta by direct thermochemical liquefaction. Fuel 74:1735–1738

Miyake J (1990) Application of photosynthetic systems for energy conversion. In: Veziroglu TN, Takahashi PK (eds) Hydrogen energy progress. VIII. Proceedings 8th WHEC. Elsevier, New York, pp 755–764

Mohan D, Pittman CU, Steele PH (2006) Pyrolysis of wood/biomass for bio-oil: a critical review. Energy Fuel 20:848–889

Mužíková Z, Pospíšil M, Šebor G (2010) The use of bioethanol as a fuel in the form of E85 fuel. Chemické listy 104(7):678–683 (in Czech)

Nichols BW (1965) Light induced changes in the lipids of *Chlorella vulgaris*. Biochim Biophys Acta 106:274–279

Ogi T, Yokoyama S, Minowa T, Dote Y (1990) Role of butanol solvent in direct liquefaction of wood. Sekiyu Gakkashi (J Japan Petr Inst) 33:383–389

Oswald WJ, Golueke C (1960) Biological transformation of solar energy. Adv Appl Microbiol 2:223–262

Pan P, Hu CW, Yang WY, Li YS, Dong LL, Zhu LF, Tong DM, Qing RW, Fan Y (2010) The direct pyrolysis and catalytic pyrolysis of *Nannochloropsis* sp residue for renewable bio-oils. Bioresour Technol 101:4593–4599

Peng WM, Wu QY, Tu PG (2000) Effects of temperature and holding time on production of renewable fuels from pyrolysis of *Chlorella protothecoides*. J Appl Phycol 12:147–152

Peng WM, Wu QY, Tu PG (2001) Pyrolytic characteristics of heterotrophic *Chlorella protothecoides* for renewable bio-fuel production. J Appl Phycol 13:5–12

Peterson AA, Vogel F, Lachance RP, Froling M, Antal MJ, Tester JW (2008) Thermochemical biofuel production in hydrothermal media: a review of sub- and supercritical water technologies. Energy Environ Sci 1:32–65

Pogaku R (2015) Advances in bioprocess technology. Springer, Cham

Pohl P, Wagner H (1972) Control of fatty acid and lipid biosynthesis in Euglena gracilis by ammonia, light and DCMU. Z Naturforsch 27:53–61

Prabandono K, Amin S (2015) Production of biomethane from marine microalgae. In: Kim SK, Lee CG (eds) Marine bioenergy: trends and developments. CRC Press/Taylor & Francis Group, Boca Raton

Pragya N, Pandey KK, Sahoo PK (2013) A review on harvesting, oil extraction and biofuels production technologies from microalgae. Renew Sust Energ Rev 24:159–171

Radakovits RRE, Jinkerson A, Darzins C (2010) Posewitz, genetic engineering of algae for enhanced biofuel production. Eukaryot Cell 9(2010):486–501

Raheem A, Wan Azlina KG, Taufiq Yap YH, Danquah MK, Harun R (2015) Thermochemical conversion of microalgal biomass for biofuel production. Renew Sust Energ Rev 49:990–999

Researchers convert algae to butanol Fuel can be used in automobiles. States News Service, March 12 2011 Issue www.newswise.com/.../researchers-convert-algae-to-butanol-fuel-can-be-used-in-auto

Rosenberg A, Gouaux J (1967) Quantitative and compositional changes in monogalactosyl and digalactosyl diglycerides during light-induced formation of chloroplasts in Euglena gracilis. J Lipid Res 8:80–83

Saifullah AZA, Karim Md A, Ahmad-Yazid A (2014) Microalgae: an alternative source of renewable energy. Am J Eng Res 3(3):330–338

Sawayama S, Inoue S, Yokoyama S (1994) Continuous culture of hydrocarbon-rich microalga Botryococcus braunii in secondarily treated sewage. Appl Microbiol Biotechnol 41:729–731

Sawayama S, Minowa T, Yokoyama SY (1999) Possibility of renewable energy production and CO2 mitigation by thermochemical liquefaction of microalgae. Biomass Bioenergy 17:33–39

Schenk PM, Thomas-Hall SR, Stephens E, Marx UC, Mussgnug JH, Posten C, Kruse O, Hankamer B (2008) Second generation biofuels: high-efficiency microalgae for biodiesel production. Bioenergy Res 1:20–43

Šebor G, Pospíšil M, Žákovec J (2006) Technical and economic analysis of suitable alternative transport fuels, research report prepared for the Ministry of Transport, ICHT Prague, June 2006. [online]. [cit. – 2012-11-09], available from: http://www.mdcr.cz/cs/Strategie/Zivotni_prostred

Sharma A, Arya SK (2017) Hydrogen from algal biomass: a review of production process. Biotechnol Rep (Amst) 14:63–69

Show PL, Tang MSY, Nagarajan D, Ling TC, Ooi CW, Chang JS (2017) A holistic approach to managing microalgae for biofuel applications. Int J Mol Sci 18:215. https://doi.org/10.3390/ijms18010215

Singh L, Kalia VC (2017) Waste biomass management – a holistic approach. Springer, Cham

Singh A, Rathore D (2017) Biohydrogen production: sustainability of current technology and future perspective. Springer, New Delhi

Soeder CJ (1986) A historical outline of applied algology. In: Richmond A (ed) Handbook of microalgal mass culture. CRC Press, Boca Raton, pp 25–41

Spoehr HA, Milner HW (1949) The chemical composition of *Chlorella*; effect of environmental conditions. Plant Physiol 24:120–149

Stucki S, Vogel F, Ludwig C, Haiduc AG, Brandenberger M (2009) Catalytic gasification of algae in supercritical water for biofuel production and carbon capture. Energy Environ Sci 2:535–541

Takac'ova A, Mackul'ak T, Smolinska M, Hut'nan M, Olejnikova P (2012) Influence of selected biowaste materials pre-treatment on their anaerobic digestion. Chem Pap 66(2):129–137

Tsukahara K, Sawayama S (2005) Liquid fuel production using microalgae. J Jpn Petr Inst 48:251–259

Vardon DR, Sharma BK, Blazina GV, Rajagopalan K, Strathmann TJ (2012) Thermochemical conversion of raw and defatted algal biomass via hydrothermal liquefaction and slow pyrolysis. Bioresour Technol 109:178–187

Varjani SJ, Agarwal AK, Gnansounou E, Gurunathan B (2018) Bioremediation: applications for environmental protection and management. Springer, Singapore

Wan YQ, Chen P, Zhang B, Yang CY, Liu YH, Lin XY, Ruan R (2009) Microwave-assisted pyrolysis of biomass: catalysts to improve product selectivity. J Anal Appl Pyrolysis 86:161–167

Wang J, Yin Y (2018) Fermentative hydrogen production using pretreated microalgal biomass as feedstock. Microb Cell Factories 17(22):1–16. https://doi.org/10.1186/s12934-018-0871-5

Wang Y, Guo W, Chen BY, Cheng CL, Lo YC, Ho SH, Chang JS, Ren N (2015) Exploring the inhibitory characteristics of acid hydrolysates upon butanol fermentation: a toxicological assessment. Bioresour Technol 198:571–576

Werner D (1966) Die Kieselsaure im Stoffwechsel von Cyclotella cryptica Reimann, Lewin and Guilard. Arch Mikrobiol 55:278–308

Yan W, Acharjee TC, Coronella CJ, Vasquez VR (2009) Thermal pretreatment of lignocellulosic biomass. Environ Prog Sustain Energy 28:435–440

Yang M (2015) The use of lignocellulosic biomass for fermentative butanol production in biorefining processes. Dissertationes Forestales. https://doi.org/10.14214/df.202

Yang YF, Feng CP, Inamori Y, Maekawa T (2004) Analysis of energy conversion characteristics in liquefaction of algae. Resour Conserv Recycl 43:21–33

Yu F, Ruan R, Steele P (2008) Consecutive reaction model for the pyrolysis of corn cob. Trans ASABE 51:1023–1028

Yu KL, Lau BF, Show PL, Ong HC, Ling TC, Chen WH, Chang JS (2017) Recent developments on algal biochar production and characterization. Bioresour Technol 246:2–11. https://doi.org/10.1016/j.biortech.2017.08.009

Zhu L (2015) Microalgal culture strategies for biofuel production: a review. Biofuels Bioprod Biorefin 9:801–814

Zhu LD, Hiltunen E, Antila E, Zhong JJ, Yuan ZH, Wang ZM (2014) Microalgal biofuels: flexible bioenergies for sustainable development. Renew Sust Energ Rev 30:1035–1046

Zou SP, Wu YL, Yang MD, Li C, Tong JM (2009) Thermochemical catalytic liquefaction of the marine microalgae *Dunaliella tertiolecta* and characterization of bio-oils. Energy Fuel 23:3753–3758

www.mdpi.com
conservancy.umn.edu
en.wikipedia.org
onlinelibrary.wiley.com
digitalcommons.usu.edu
microbewiki.kenyon.edu
www.fona.de
www.ncbi.nlm.nih.gov
www.yesitekhob.com
econpapers.repec.org
eprints.qut.edu.au
www.assb.pl
d-nb.info
pdfs.semanticscholar.org
archive.org

Chapter 8
Current Trends and the Future of the Algae-Based Biofuels Industry

Abstract Current trends and the future of the algae-based biofuels industry are discussed. Some of the leading companies involved with third generation biofuel research and development are also presented.

Keywords Biofuels · Algae · Third generation biofuel · Algal fuel · Transport fuel · Biodiesel · Bioethanol · Algal oil · Hydrogen · Aviation fuel

Biofuels from algae is an immensely important subject strategically and is also controversial. Interest in this area is global, both the developed nations and the emerging economies are interested. Algal fuels are not commercial yet, but their economic outlook is promising (Chisti 2007, 2008a, b; Stephens et al. 2010a, b; Norsker et al. 2011). Many important scientific and technical barriers however, remain to be overcome before the large-scale production of microalgal fuels can become a commercial reality. Technological developments, including advances in photobioreactor design, biomass harvesting, drying, and processing are the important areas leading to increased cost-effectiveness and hence, effective commercial implementation of the biofuel from microalgae strategy.

Several startup companies are attempting to commercialize algal fuels (Table 8.1). There already are many big players in the business of generation of biofuels from microalgae in the USA i.e. Solazyme, Sapphire Energy, PetroSun, Joule Unlimited, Green Fuel Technologies Corporation, Global Green Algae, Gevo, Algenol. In Europe, there are: Alpha Biotech (France), Algae-farms (Greece), Powerfuel.de (Germany), Algae Link (Spain), and Varicon Aqua Solutions Ltd. and British Algoil Ltd. (UK). All these companies are already producing commercial scale biodiesel, bioethanol, algal oil, hydrogen, and aviation fuel from algae. Extensive research is going on in this field, but relaxing the legislation with respect to growing genetically modified algae in open ponds and attracting more innovative projects is the requirement of this field. Coming up with vigorous and cost competitive conversion technologies would be staggeringly advantageous to the biofuel industry and to humankind in the long run.

Table 8.1 Companies involved in commercialization of algal biofuels

Company	Website
Algenol Biofuels, Bonita Springs, FL, USA	www.algenolbiofuels.com
Aquaflow Nelson, New Zealand	www.aquaflowgroup.com
Aurora Algae, Inc. Hayward CA, US	www.aurorainc.com
Bioalgene Hayward, CA, USA	www.bioalgene.com
Bionavitas, Inc. Hayward, CA, USA	www.bionavitas.com
Bodega Algae, LLC Boston, MA, USA	www.bodegaalgae.com
LiveFuels, Inc. San Carlos, CA, USA	www.livefuels.com
PetroAlgae Inc. Melbourne, FL, USA	www.petroalgae.com
Phyco Biosciences Chandler, AZ, USA	www.phyco.net
Sapphire Energy, Inc. San Diego, CA, USA	www.sapphireenergy.com
Seambiotic Ltd. Tel Aviv, Israel	www.seambiotic.com
Solazyme, Inc. South San Francisco, CA, USA	www.solazyme.com
Solazyme, Inc. South San Francisco, CA, USA	www.solazyme.com
Solix Biofuels, Inc. Fort Collins, CO, USA	www.solixbiofuels.com
Synthetic Genomics Inc. La Jolla, CA, USA	www.syntheticgenomics.com

Based on Chisti and Yan (2011)

Notwithstanding the many odds, crude oil from algae will likely be an important energy feedstock of the future (Stephens et al. 2010a; Chisti 2010; Tredici 2010; Greenwell et al. 2010). Algal oil could be converted into diesel, gasoline and jet fuel, and become a renewable feedstock for making plastics and the other chemicals that are now obtained from petroleum at great cost to the environment (Chisti 2007, 2010). Displacing petroleum derived transport fuels with fuels from algae could potentially reduce emission of carbon dioxide by about 30% in the USA. Production of liquid fuels from algae is of course already technically possible, but expensive compared to petroleum fuels. A major obstacle to investment in fuels from algae technologies is the susceptibility of petroleum price to unpredictable and large fluctuations. Algal oil is likely to be economically viable in a scenario with crude petroleum selling for P$100 per barrel (Stephens et al. 2010b).

The concept of biofuels from algae is conceptually fundamentally sound (Stephens et al. 2010a; Chisti 2007, 2008a, b, 2010; Weyer et al. 2010). Despite the existing poorly developed production methods, algal fuels achieve a net positive energy recovery (Chisti 2008a, b; Weyer et al. 2010; Xu et al. 2011) but how much exactly, remains controversial. Water footprint of algal biodiesel appears to be smaller than the water footprints of biodiesel from other crops (Yang et al. 2011). Of the two major types of large-scale algae culture systems (Chisti 2007; Terry and Raymond 1985; Carvalho et al. 2006) open ponds have a low productivity compared to photobioreactors (Chisti 2007, 2008a, b). Photobioreactors require a high initial capital investment, but appear to be able to produce biomass at a reduced price (Chisti 2007; Norsker et al. 2011) at least in some cases. In addition, photobioreactors produce a much more concentrated algal broth than the open ponds and this reduces the dewatering costs significantly. Using tubular types of photobioreactors, it may be possible to produce dewatered algal biomass at around €4 per kilogram dry weight (Acien Fernandez et al. 2001; Molina Grima et al. 2000; Sanchez Miron et al. 2003; Norsker et al. 2011). Other studies have reached similar conclusions (Chisti 2007). As the scale of the production facility is increased, the unit cost of producing the algal biomass will reduce further. In the longer term, genetic engineering will likely have the greatest effect on the feasibility of algal biofuels. Advances in algal biomass separation method from the water and extraction of the oil from the biomass, will improve the prospects of algal oil (Stephens et al. 2010a; Chisti 2010). With genetic engineering, the cells of certain photosynthetic microorganisms have been coaxed into secreting the oil which would normally be retained within the cell (Anonymous 2010), thus making the oil recovery simpler. Algal species engineered to use atmospheric nitrogen instead of nitrogen fertilizers that are now needed, will be a significant step forward, as production of nitrogen fertilizers is heavily dependent of petroleum (Chisti 2010). Fuels from algae certainly look promising. They may already be viewed as competitive with petroleum fuels, if the full environmental impact of the latter types of fuels is taken into consideration. Climate change issues may force us to move beyond petroleum long before it is exhausted.

Biofuels remain the most environment friendly and practical solution to the global fuel crisis.

There are a significant number of economic and technical challenges associated with the usage of microalgae in the biofuels industry. Harvesting microalgae is a major problem. The unicellular algae that stores lipids have low densities and are located in suspensions making separations quite laborious. The extraction processes used for large-scale production are particularly complex and are still in the early development stages (Rawat et al. 2013). Microalgae cultivated in open pond systems are prone to contamination. Bacterial contamination aggressively competes for nutrients and oxidises the organic matter, which can lead to culture putrefaction. They are also susceptible to protozoa and zooplankton grazers that consume microalgae and may destroy the concentrations of algae in a short time (Rawat et al. 2013). In open pond systems there is also loss of water through evaporation and in order to maintain a fixed volume and salinity in the culture it is important to add large quantities of freshwater (Das et al. 2015). Other challenges that inhibit the commercialization of algal based biofuel production include; difficulties in finding rapid growing algae strains with high oil content, photosynthetic efficiency, simple algae culture harvesting systems, infrastructure, operation and maintenance costs and the ability to develop economical photo-bioreactor designs (Adenle et al. 2013).

Microalgae have the potential to be used to produce certain biofuels without the controversial issues associated with land use, the environment and sustainability. There is lot of focus on the possibility of thermal conversion using hydrothermal liquefaction to transform microalgae biomass into usable biofuels. The feasibility of up-scaling a microalgae cultivation system requires testing particularly in terms of economic viability and product yield. The use of microalgae in biotechnology certainly has the possibility to transform the field, this potential increases with the use of transgenic algal strains (Rosenberg et al. 2008). The success stories with respect to synthetic biology and genetically engineering microalgae show a bright future for the biofuels industry.

References

Acien Fernandez FG, Fernandez Sevilla JM, Sanchez Perez JA, Molina Grima E, Chisti Y (2001) Airlift-driven external-loop tubular photobioreactors for outdoor production of microalgae: assessment of design and performance. Chem Eng Sci 56:2721–2732

Adenle AA, Haslam GE, Lee L (2013) Global assessment of research and development for algae biofuel production and its potential role for sustainable development in developing countries. Energy Policy 61:182–195

Anonymous (2010) Reengineered microorganisms leach renewable biofuel. Chem Eng Prog, May 13, 2010

Carvalho AP, Meireles LA, Malcata FX (2006) Microalgal reactors: a review of enclosed system designs and performances. Biotechnol Prog 22:1490–1506

Chisti Y (2007) Biodiesel from microalgae. Biotechnol Adv 25:294–306

Chisti Y (2008a) Biodiesel from microalgae beats bioethanol. Trends Biotechnol 26:126–131

Chisti Y (2008b) Response to Reijnders: do biofuels from microalgae beat biofuels from terrestrial plants? Trends Biotechnol 26:351–352

Chisti Y (2010) Fuels from microalgae. Biofuels 1:233–235

Chisti Y, Yan J (2011) Algal biofuels – a status report. Appl Energy 88:3277–3279

Das P, Thaher MI, Hakim MAQMA, Al-Jabri HMSJ (2015) Sustainable production of toxin free marine microalgae biomass as fish feed in large scale open system in the Qatari desert. Bioresour Technol 192:97–104

Greenwell HC, Laurens LML, Shields RJ, Lovitt RW, Flynn KJ (2010) Placing microalgae on the biofuels priority list: a review of the technological challenges. J Roy Soc Interface 7:703–726

Molina Grima E, Acien Fernandez FG, Garcia Camacho F, Camacho Rubio F, Chisti Y (2000) Scale-up of tubular photobioreactors. J Appl Phycol 12:355–368

Norsker N-H, Barbosa MJ, Vermue MH, Wijffels RH (2011) Microalgal production – a close look at the economics. Biotechnol Adv 29:24–27

Rawat I, Kumar RR, Mutanda T, Bux F (2013) Biodiesel from microalgae: a critical evaluation from laboratory to large scale production. Appl Energy 103:444–467

Rosenberg JN, Oyler GA, Wilkinson L, Betenbaugh MJ (2008) A green light for engineered algae: redirecting metabolism to fuel a biotechnology revolution. Curr Opin Biotechnol 19(5):430–436

Sanchez Miron A, Ceron Garcia M-C, Contreras Gomez A, Garcia Camacho F, Molina Grima E, Chisti Y (2003) Shear stress tolerance and biochemical characterization of Phaeodactylum tricornutum in quasi steady-state continuous culture in outdoor photobioreactors. Biochem Eng J 16:287–297

Stephens E, Ross IL, King Z, Mussgnug JH, Kruse O, Posten C (2010a) An economic and technical evaluation of microalgal biofuels. Nat Biotechnol 28:126–128

Stephens E, Ross IL, Mussgnug JH, Wagner LD, Borowitzka MA, Posten C, Kruse O and , Hankamer B (2010b). Future prospects of microalgal biofuel production systems. Trends Plant Sci;15:554–564

Terry KL, Raymond LP (1985) System design for the autotrophic production of microalgae. Enzym Microb Technol 7:474–487

Tredici MR (2010) Photobiology of microalgae mass cultures: understanding the tools for the next green revolution. Biofuels 1:143–162

Weyer KM, Bush DR, Darzins A, Willson BD (2010) Theoretical maximum algal oil production. Bioenergy Res 3:204–213

Xu L, Brilman DWF, Withag JAM, Brem G, Kersten S (2011) Assessment of a dry and a wet route for the production of biofuels from microalgae: energy balance analysis. Bioresour Technol 102:5113–5122

Yang J, Xu M, Zhang X, Hu Q, Sommerfeld M, Chen Y (2011) Life-cycle analysis on bio-diesel production from microalgae: water footprint and nutrients balance. Bioresour Technol 102:159–165

Index

Printed in the United States
By Bookmasters